Remote Sensing in Applications of Geoinformation

Remote Sensing in Applications of Geoinformation

Editor

Silas Michaelides

MDPI • Basel • Beijing • Wuhan • Barcelona • Belgrade • Manchester • Tokyo • Cluj • Tianjin

Editor
Silas Michaelides
Eratosthenes Centre of Excellence
Cyprus University of Technology
Limassol
Cyprus

Editorial Office
MDPI
St. Alban-Anlage 66
4052 Basel, Switzerland

This is a reprint of articles from the Special Issue published online in the open access journal *Remote Sensing* (ISSN 2072-4292) (available at: https://www.mdpi.com/journal/remotesensing/special_issues/rs_geoinf).

For citation purposes, cite each article independently as indicated on the article page online and as indicated below:

LastName, A.A.; LastName, B.B.; LastName, C.C. Article Title. *Journal Name* **Year**, *Volume Number*, Page Range.

ISBN 978-3-0365-2325-5 (Hbk)
ISBN 978-3-0365-2326-2 (PDF)

© 2021 by the authors. Articles in this book are Open Access and distributed under the Creative Commons Attribution (CC BY) license, which allows users to download, copy and build upon published articles, as long as the author and publisher are properly credited, which ensures maximum dissemination and a wider impact of our publications.

The book as a whole is distributed by MDPI under the terms and conditions of the Creative Commons license CC BY-NC-ND.

Contents

About the Editor vii

Silas Michaelides
Editorial for Special Issue "Remote Sensing in Applications of Geoinformation"
Reprinted from: *Remote Sens.* **2021**, *13*, 33, doi:10.3390/rs13010033 1

Elsayed Said Mohamed, A. A El Baroudy, T. El-beshbeshy, M. Emam, A. A. Belal, Abdelaziz Elfadaly, Ali A. Aldosari, Abdelraouf. M. Ali and Rosa Lasaponara
Vis-NIR Spectroscopy and Satellite Landsat-8 OLI Data to Map Soil Nutrients in Arid Conditions: A Case Study of the Northwest Coast of Egypt
Reprinted from: *Remote Sens.* **2020**, *12*, 3716, doi:10.3390/rs12223716 5

Kyriacos Themistocleous, Christiana Papoutsa, Silas Michaelides and Diofantos Hadjimitsis
Investigating Detection of Floating Plastic Litter from Space Using Sentinel-2 Imagery
Reprinted from: *Remote Sens.* **2020**, *12*, 2648, doi:10.3390/rs12162648 25

Lucille Alonso and Florent Renard
A New Approach for Understanding Urban Microclimate by Integrating Complementary Predictors at Different Scales in Regression and Machine Learning Models
Reprinted from: *Remote Sens.* **2020**, *12*, 2434, doi:10.3390/rs12152434 43

Eleni Kokinou and Costas Panagiotakis
Automatic Pattern Recognition of Tectonic Lineaments in Seafloor Morphology to Contribute in the Structural Analysis of Potentially Hydrocarbon-Rich Areas
Reprinted from: *Remote Sens.* **2020**, *12*, 1538, doi:10.3390/rs12101538 79

Elena Barbierato, Iacopo Bernetti, Irene Capecchi and Claudio Saragosa
Integrating Remote Sensing and Street View Images to Quantify Urban Forest Ecosystem Services
Reprinted from: *Remote Sens.* .**2020**, *12*, 329, doi:10.3390/rs12020329 97

Thomas Dimopoulos, Nikolaos P. Bakas
Sensitivity Analysis of Machine Learning Models for the Mass Appraisal of Real Estate. Case Study of Residential Units in Nicosia, Cyprus
Reprinted from: *Remote Sens.* **2019**, *11*, 3047, doi:10.3390/rs11243047 119

Georgios A. Kordelas, Ioannis Manakos, Gaëtan Lefebvre and Brigitte Poulin
Automatic Inundation Mapping Using Sentinel-2 Data Applicable to Both Camargue and Doñana Biosphere Reserves
Reprinted from: *Remote Sens.* **2019**, *11*, 2251, doi:10.3390/rs11192251 135

Iñaki Prieto, Jose Luis Izkara and Elena Usobiaga
The Application of LiDAR Data for the Solar Potential Analysis Based on Urban 3D Model
Reprinted from: *Remote Sens.* **2019**, *11*, 2348, doi:10.3390/rs11202348 155

About the Editor

Silas Michaelides is currently affiliated with the Eratosthenes Centre of Excellence of the Cyprus University of Technology. He has been the Director of the Department of Meteorology of Cyprus, a position he has reached having climbed through the scientific ranks of this governmental organization for more than 40 years. He holds a PhD in Meteorology, an MSc in Agricultural Meteorology, a Master's degree in Public Sector Management, and a BSc in Mathematics. He has published 125 papers in peer-reviewed scientific journals, several of which are on the remote sensing of precipitation. He has also published one book on precipitation, and he has been the Guest Editor for more than 30 Special Issues and Conference Proceedings. He is a Member Emeritus of the European Geosciences Union, a Member of the American Meteorological Society, a Fellow of the Royal Meteorological Society and a Member of the Hellenic Meteorological Society. He is currently the Vice Chairman of the Cyprus Remote Sensing Society.

Editorial

Editorial for Special Issue "Remote Sensing in Applications of Geoinformation"

Silas Michaelides [1,2]

1 ERATOSTHENES Centre of Excellence, 3117 Limassol, Cyprus; silas.michaelides@cut.ac.cy; Tel.: +357-99493072
2 Department of Civil Engineering and Geomatics, Cyprus University of Technology, 3036 Limassol, Cyprus

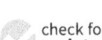

Citation: Michaelides, S. Editorial for Special Issue "Remote Sensing in Applications of Geoinformation". *Remote Sens.* 2021, 13, 33. https://dx.doi.org/10.3390/rs13010033

Received: 15 December 2020
Accepted: 20 December 2020
Published: 23 December 2020

Publisher's Note: MDPI stays neutral with regard to jurisdictional claims in published maps and institutional affiliations.

Copyright: © 2020 by the author. Licensee MDPI, Basel, Switzerland. This article is an open access article distributed under the terms and conditions of the Creative Commons Attribution (CC BY) license (https://creativecommons.org/licenses/by/4.0/).

1. Introduction

The diffusion of knowledge and information is currently more forceful than ever. Indeed, we are witnessing the enormous transformative power of the knowledge revolution that our societies, industries and economies are subject to. One of the drivers in the current knowledge-based society is remote sensing which is commonly defined as the acquisition of information about an object without making physical contact with it. In a more restricted sense, remote sensing refers to the science and technology of acquiring information about the Earth's surface. Remote sensing delivers a wealth of information which would otherwise be inconceivable. Geoinformatics is defined as the scientific discipline for the acquisition, storage, analysis and presentation of geospatial information.

Geoinformation is a field that greatly benefits from the technological advances in remote sensing. The numerous advantages of using remote sensing in geoinformation are demonstrated by the large number of application-oriented endeavors already undertaken. Depending on the need (i.e., scientific, societal, mapping, planning, hazard mitigation, etc.), emphasis may be placed on different facets of geoinformation.

This Special Issue of Remote Sensing comprises a contribution to the multi-faceted range of applications of remote sensing in geoinformation. It hosts eight papers focusing on a broad range of scientific contributions underscoring this synergetic approach to remote sensing and geoinformation. These papers were selected from the presentations at the "7th International Conference on Remote Sensing and Geoinformation of the Environment (RSCy2019)" held in Paphos, Cyprus, from 18 to 21 March 2019.

The next section summarizes the individual articles hosted in this Special Issue entitled "Remote Sensing in Applications of Geoinformation". The articles are presented in alphabetical order based on the first author's name.

2. Overview of Contributions

The study by Alonso and Renard [1] proposes modeling air temperatures, measured during four mobile campaigns carried out during the summer months, between 2016 and 2019, in Lyon (France), in clear-sky weather. The study proposes the usage of regression models based on 33 explanatory variables from traditionally used data, namely, from remote sensing by LiDAR (light detection and ranging) or Landsat 8 satellite acquisition. Three types of statistical regressions were explored: partial least square regression, multiple linear regression and random forest regression. The authors have shown that variables such as surface temperature, normalized difference vegetation index and modified normalized difference water index have a strong impact on the estimation model. This study contributes to the emergence of urban cooling systems.

The aim of the study by Barbierato et al. [2] is to create a general-purpose set of ecological metrics by combining remote sensing and proximate sensing (Street View) approaches with data retrieved from Google Street View, to quantify urban forest ecosystem services and provide a widely transferable methodology. In this respect, remote sensing metrics were calculated by combining high-resolution multispectral images and LiDAR data to produce indices at different altitudes with respect to the ground. The ecological metrics from proximate sensing were then calculated by semantic segmentation using pretrained deep segmentation neural networks. To estimate the validity of this approach, a set of ecological metrics was used to classify contiguous homogeneous areas of a city through a spatial clustering algorithm.

Dimopoulos and Bakas [3] investigate how complex machine learning models work, regarding real estate price predictions, and present the various models and the corresponding results. They explain the analyzed dataset as well as its variables. The machine learning methods utilized for the target task are presented, as well as the generic algorithm to obtain the closed-form formula for the higher order regression model, via an automated, stepwise method. They also present the sensitivity analysis results of the predictors, regarding real estate prices; the influence of the dataset volume is also investigated, by a parametric study, for a variety of partitions of the given dataset. Important constraints have been identified, such as the transparency of models and the repeatability of the results.

Kokinou and Panagiotakis [4] present novel pattern recognition techniques applied to bathymetric data from two large areas in the Eastern Mediterranean. Their objectives are: (a) to demonstrate the efficiency of this methodology, (b) to highlight the quick and accurate detection of both hydrocarbon related tectonic lineaments and salt structures affecting seafloor morphology and (c) to reveal new structural data in areas poised for hydrocarbon exploration. In this work, they first apply a multiple filtering and sequential skeletonization scheme inspired by the hysteresis thresholding technique. Subsequently, they categorize each linear and curvilinear segment on the seafloor skeleton (medial axis) based on the strength of detection as well as the length, direction and spatial distribution. Finally, they compare the seafloor skeleton with ground truth data.

The study by Kordelas et al. [5] examines the applicability of a novel automatic local thresholding unsupervised methodology for separating inundated areas from non-inundated ones and proposes alternatives to the original approach to enhance accuracy and applicability for both Camargue (France) and Doñana (Spain) wetlands. Each examined alternative approach relies on a specific band or band combination, acknowledged as effective by the underlying physics, and a specific approach for estimating splitting thresholds. The different Sentinel-2 based inputs examined for estimating thresholds include: (a) Band 11 (SWIR-1); (b) product of Band 12 (SWIR-2) and Band 8A (NIR); and (c) product of SWIR-1 and NIR (near infra red). The different methods for estimating splitting thresholds include: (a) minimum entropy thresholding and (b) Otsu's algorithm. The results of the alternative approaches are compared against reference maps, provided for Doñana and Camargue by local research institutes, based on locally developed water detection models.

The mapping of soil nutrients is a key issue for numerous applications and research fields ranging from global change to environmental degradation and from sustainable soil management to the precision agriculture concept. The characterization, modeling and mapping of soil properties at diverse spatial and temporal scales are key factors required for different environments. The paper by Mohamed et al. [6] focuses on the use and comparison of soil chemical analyses, visible near infrared and shortwave infrared spectroscopy, partial least-squares regression, ordinary Kriging, and Landsat-8 operational land imager images, to inexpensively analyze and predict the content of different soil nutrients (nitrogen (N), phosphorus (P) and potassium (K)), pH and soil organic matter in arid conditions. To achieve this aim, 100 surface samples of soil were gathered to a depth of 25 cm in the Wadi El-Garawla area (northwest coast of Egypt) and chemical analyses and reflectance spectroscopy in the wavelength range from 350 to 2500 nm was utilized.

Solar maps are becoming a popular resource and are available via the web to help plan investments for the benefits of renewable energy. These maps are especially useful when the results have high accuracy. LiDAR technology currently offers high-resolution data sources that are very suitable for obtaining an urban 3D geometry with high precision. Three dimensional visualization also offers a more accurate and intuitive perspective of reality than 2D maps. The paper by Prieto et al. [7] presents a new method for the calculation and visualization of the solar potential of building roofs in an urban 3D model, based on LiDAR data. The paper describes the proposed methodology to (a) calculate the solar potential, (b) generate an urban 3D model, (c) semanticize the urban 3D model with different existing and calculated data and (d) visualize the urban 3D model in a 3D web environment. The paper presents the workflow and results of application to the city of Vitoria-Gasteiz in Spain.

Themistocleous et al. [8] conducted a study to determine if plastic targets on the sea surface can be detected using remote sensing techniques with Sentinel-2 data. A target made up of plastic water bottles with a surface measuring 3 m × 10 m was created and was subsequently placed in the sea near the Old Port in Limassol, Cyprus. An unmanned aerial vehicle (UAV) was used to acquire multispectral aerial images of the area of interest during the same time as the Sentinel-2 satellite overpass. Spectral signatures of the water and the plastic litter after it was placed in the water were taken with a Spectra Vista Corporation HR1024 spectroradiometer. The study found that the plastic litter target was easiest to detect in the NIR wavelengths. Seven established indices for satellite image processing were examined to determine whether they can identify plastic litter in the water. Further, the authors examined two new indices, the plastics index and the reversed normalized difference vegetation index to be used in the processing of the satellite image. The proposed plastic index was able to identify plastic objects floating on the water surface and was the most effective index in identifying the plastic litter target in the sea.

3. Conclusions

The scientific contributions in this Special Issue aim at informing and updating the scientific communities involved in geoinformation and remote sensing on findings in important areas of remote sensing in applications of geoinformation. Remote sensing and geoinformation technologies have a pivotal role in innovation; they also offer solutions to major environmental issues and contribute to the modernization of many scientific developments, with a significant impact on the quality of life and the economy.

Remote sensing has long been proven to be a valuable tool in a wide range of disciplines for the study of the environment, such as, weather, monitoring of air pollution, the environmental control and management, mapping of geomorphological structures and the prevention and mitigation of natural disasters, etc. Remote sensing has also found fertile ground in the field of geoinformation, as is very aptly indicated by the examples in this volume.

On the one hand, the technological advances in remote sensing are proliferating at a fast pace. On the other hand, the evolving field of geoinformation is increasingly becoming a societal commodity. Fusion of remote sensing and geoinformatics opens new challenging routes for further investigations, research and experimentation. By presenting state-of-the-art data sources, technologies and methodologies, this Special Issue aspires to stimulate further research in the increasingly expanding field of applications of remote sensing in geoinformation.

Funding: Silas Michaelides was supported by the EXCELSIOR project (www.excelsior2020.eu) that has received funding from the European Union's Horizon 2020 Research and Innovation Programme, under grant agreement no. 857510, as well as matching co-funding by the Government of the Republic of Cyprus through the Directorate General for the European Programmes, Coordination and Development.

Institutional Review Board Statement: Not applicable.

Informed Consent Statement: Not applicable.

Data Availability Statement: Not applicable.

Acknowledgments: As the Guest Editor of this Special Issue entitled "Remote Sensing in Applications of Geoinformation", I would like first to thank the authors who have submitted papers to this volume and for sharing their scientific findings. All the authors collaborated closely with the

Guest Editor and with the Editorial Office in the effort to achieve the highest possible quality of the present volume. The role of the eminent reviewers is greatly appreciated; their timely and thorough reviews added value to the scientific quality of the volume. Last but not least, I wish to express my gratitude to the editorial staff of Remote Sensing for their collaboration and prompt efforts in completing this task.

Conflicts of Interest: The author declares no conflict of interest.

References

1. Alonso, L.; Renard, F. A New Approach for Understanding Urban Microclimate by Integrating Complementary Predictors at Different Scales in Regression and Machine Learning Models. *Remote Sens.* **2020**, *12*, 2434. [CrossRef]
2. Barbierato, E.; Bernetti, I.; Capecchi, I.; Saragosa, C. Integrating Remote Sensing and Street View Images to Quantify Urban Forest Ecosystem Services. *Remote Sens.* **2020**, *12*, 329. [CrossRef]
3. Dimopoulos, T.; Bakas, N. Sensitivity Analysis of Machine Learning Models for the Mass Appraisal of Real Estate. Case Study of Residential Units in Nicosia, Cyprus. *Remote Sens.* **2019**, *11*, 3047. [CrossRef]
4. Kokinou, E.; Panagiotakis, C. Automatic Pattern Recognition of Tectonic Lineaments in Seafloor Morphology to Contribute in the Structural Analysis of Potentially Hydrocarbon-Rich Areas. *Remote Sens.* **2020**, *12*, 1538. [CrossRef]
5. Kordelas, G.A.; Manakos, I.; Lefebvre, G.; Poulin, B. Automatic Inundation Mapping Using Sentinel-2 Data Applicable to Both Camargue and Do'ana Biosphere Reserves. *Remote Sens.* **2019**, *11*, 2251. [CrossRef]
6. Mohamed, E.S.; El Baroudy, A.A.; El-beshbeshy, T.; Emam, M.; Belal, A.A.; Elfadaly, A.; Aldosari, A.A.; Ali, A.M.; Lasaponara, R. Vis-nir Spectroscopy and Satellite Landsat-8 Oli Data to Map Soil Nutrients in Arid Conditions: A Case Study of the Northwest Coast of Egypt. *Remote Sens.* **2020**, *12*, 3716. [CrossRef]
7. Prieto, I.; Izkara, J.L.; Usobiaga, E. The Application of LiDAR Data for the Solar Potential Analysis Based on Urban 3D Model. *Remote Sens.* **2019**, *11*, 2348. [CrossRef]
8. Themistocleous, K.; Papoutsa, C.; Michaelides, S.; Hadjimitsis, D. Investigating Detection of Floating Plastic Litter from Space Using Sentinel-2 Imagery. *Remote Sens.* **2020**, *12*, 2648. [CrossRef]

Article

Vis-NIR Spectroscopy and Satellite Landsat-8 OLI Data to Map Soil Nutrients in Arid Conditions: A Case Study of the Northwest Coast of Egypt

Elsayed Said Mohamed [1], A. A El Baroudy [2], T. El-beshbeshy [2], M. Emam [1], A. A. Belal [1], Abdelaziz Elfadaly [1,3], Ali A. Aldosari [4], Abdelraouf. M. Ali [1,5] and Rosa Lasaponara [3,*]

1. National Authority for Remote Sensing and Space Sciences, Cairo 1564, Egypt; elsayed.salama@narss.sci.eg (E.S.M.); emam@narss.sci.eg (M.E.); massud-am@rudn.ru (A.A.B.); abdelaziz.elfadaly@narss.sci.eg (A.E.); abdelraouf.ali@narss.sci.eg (A.M.A.)
2. Soils and Water Department, Faculty of Agriculture, Tanta University, Gharbiya 31527, Egypt; drbaroudy@agr.tanta.edu.eg (A.A.E.B.); elbeshbeshy@agr.tanta.edu.eg (T.E.-b.)
3. Italian National Research Council, C.da Santa Loja, Tito Scalo, 85050 Potenza, Italy
4. Geography Department, King Saud University, Riyadh 11451, Saudi Arabia; adosari@ksu.edu.sa
5. Agrarian-Technological Institute of the Peoples' Friendship University of Russia, ul. Miklukho-Maklaya 6, 117198 Moscow, Russia
* Correspondence: rosa.lasaponara@imaa.cnr.it; Tel.: +39-327-709-0396

Received: 12 September 2020; Accepted: 1 November 2020; Published: 12 November 2020

Abstract: The mapping of soil nutrients is a key issue for numerous applications and research fields ranging from global changes to environmental degradation, from sustainable soil management to the precision agriculture concept. The characterization, modeling and mapping of soil properties at diverse spatial and temporal scales are key factors required for different environments. This paper is focused on the use and comparison of soil chemical analyses, Visible near infrared and shortwave infrared VNIR-SWIR spectroscopy, partial least-squares regression (PLSR), Ordinary Kriging (OK), and Landsat-8 operational land imager (OLI) images, to inexpensively analyze and predict the content of different soil nutrients (nitrogen (N), phosphorus (P), and potassium (K)), pH, and soil organic matter (SOM) in arid conditions. To achieve this aim, 100 surface samples of soil were gathered to a depth of 25 cm in the Wadi El-Garawla area (the northwest coast of Egypt) using chemical analyses and reflectance spectroscopy in the wavelength range from 350 to 2500 nm. PLSR was used firstly to model the relationship between the averaged values from the ASD spectroradiometer and the available N, P, and K, pH and SOM contents in soils in order to map the predicted value using Ordinary Kriging (OK) and secondly to retrieve N, P, K, pH, and SOM values from OLI images. Thirty soil samples were selected to verify the validity of the results. The randomly selected samples included the spatial diversity and characteristics of the study area. The prediction of available of N, P, K pH and SOM in soils using VNIR-SWIR spectroscopy showed high performance (where R^2 was 0.89, 0.72, 0.91, 0.65, and 0.75, respectively) and quite satisfactory results from Landsat-8 OLI images (correlation R^2 values 0.71, 0.68, 0.55, 0.62 and 0.7, respectively). The results showed that about 84% of the soils of Wadi El-Garawla are characterized by low-to-moderate fertility, while about 16% of the area is characterized by high soil fertility.

Keywords: soil nutrients; field spectroscopy; Landsat (OLI); partial least-squares and regression; Wadi El-Garawla

1. Introduction

Soil is a very complex ecosystem made up of biotic and abiotic factors that strongly differ from one environment to another. The characterization, modelling and mapping of soil properties are key factors for implementing good agricultural management practices [1–4] to maintain ecological balances and prevent land degradation in arid and semiarid environments. As an example, the accumulation of salts and soil nutrients in arid conditions is affected by many factors, such as topography, geology, climate, soil moisture, land use, agricultural activity, and local environmental conditions [5–8]. The traditional methods for estimating different soil properties typically involves extensive field work and laboratory analysis and, therefore, are not only expensive and time consuming but also may be affected by significant uncertainty. Therefore, over the last four decades, to model and map soil properties in a cost-effective manner at various scales, remotely sensed imagery has been proposed and used in combination with field measurements [9–12]. Important soil properties such as salinity, texture, minerals, and organic matter have been successfully characterized and investigated using multispectral scanner (MS), Landsat-8 operational land imager (OLI), Landsat-5 thematic mapper (TM), Landsat-7 enhanced thematic mapper plus (ETM+) [13].

Over the past two decades, scientists throughout the world have focused their interest on new technologies such as the visible–near-infrared (Vis-NIR) spectroscopy to identify and characterize soil in terms of (but not only) clay mineralogy, soil organic matter (SOM), soil composition, and soil texture [14–17]. It is well recognized that the absorption spectrum in the NIR zone (780–2500 nm) can be used for estimating H_2O, CO_2, OH, SO_4, and CO_3 groups [18]; furthermore, soil nutrients can be identified using NIR spectroscopy, particularly for estimating N, K, and P soil content (with expected satisfactory coefficients of correlation around 0.72 and 0.68 for N and K, respectively), and with higher value in the case of phosphorus (around 0.84) [19]. Moreover, additional independent studies have shown that calcium, potassium, magnesium, and sodium can be predicted using statistical models such as partial least-squares regression (PLSR) [20,21]. As a whole, today, Vis-NIR techniques are recognized to be effective for the quantitative retrieval of soil characteristics and usually provide good indications of soil quality [22]. Nevertheless, some critical issues have still to be faced, such as, for example, the estimation of carbonate and the gypsum contents that is still today a controversial issue. In fact, some studies highlighted that spectroscopic techniques cannot suitably predict the carbonate content (correlation lower than 0.52), whereas other studies pointed out that the joint use of spectroscopic techniques and PLSR improved the estimation with correlation values ranging from 0.86 to 0.91 [23–25].

From the methodological point of view, analytical methods based on changes in specific reflectance (in the visible range from 400 to 700 nm, and in the near-infrared range from 700 to 2500 nm [26,27]), enable the discrimination of different soil properties, such as pH, organic carbon, electrical conductivity, texture, nitrate–nitrogen, available phosphorus, exchangeable potassium, cation exchange capacity, exchangeable calcium, and exchangeable aluminum. Moreover, several prediction models have been used to assess soil properties based on reflectance spectroscopy, such as artificial neural networks (ANN), partial least square regression (PLSR), stepwise multiple linear regression (SIMR), multivariate adaptive regression splines (MARS), locally weighted regression (LWR), and principal components regression (PCR) [18,28].

As a whole, today, one of the major challenges to be faced is the need to develop low-cost methods for mapping soil properties over large areas and, on the other hand, it is important to consider that agricultural management needs a rapid analysis to identify the deficiency of elements in the soil and crops. To cope with this issue, Vis-NIR reflectance spectroscopy coupled with satellite data can suitably complement in situ analyses [29,30]. The timely availability of quantitative information on soil properties and their spatial distribution is extremely relevant for sustainable agricultural to achieve development, reducing the negative effects on soil and environment [31–33]. This is extremely important in arid and semi-arid areas which have several limiting factors for soil fertility, such as low nitrogen, phosphorus, scarcity of irrigation water, and low soil organic matter. Moreover, the mapping soil properties and fertility provides good indicators of land degradation [34–36] and/or evidence of land capacity.

An effort in this context is made in this paper, which is focused on the evaluation of soil nutrients (N, P, K), SOM, and pH in the arid area of Wadi El-Garawla (the northwest coast of Egypt) jointly using chemical analyses, Vis-NIR spectroscopy and satellite Landsat-8 data that are freely available from the NASA web site. In detail, the PLSR was used firstly to model the relationship between the averaged values from the analytical spectral devices (ASD) spectroradiometer and the soil's available nitrogen (N), phosphorus (P), and potassium (K) content, along with the pH and the soil organic matter (SOM); and (ii) secondly to retrieve N, P, K, pH, and SOM values from the OLI images. Thirty soil samples were selected to verify the validity of the results. The randomly selected samples included the spatial diversity and characteristics of the study area.

The approach herein proposed enabled us to (I) model the relationship between the Spectral reflectance by ASD spectroradiometer and the laboratory analysis of soil of soil properties (N, P, K), pH, and SOM; (II) map the predicted soil properties using OK; (III) map the predicted soil properties from Landsat OLI images; and (IV) map the soil fertility status.

Today the availability of open satellite data from national and international space agencies strongly facilitates the investigation of soil properties, and their timely availability enables a prompt update and spatial distribution over a large area as necessary to support soil management strategies and to update information on the input parameters of crop models.

2. Materials and Methods

2.1. Experimental Site

The investigated area is located on the northwestern side of the coastal zone in the western desert area of Egypt. Wadi El-Garawla is located about 18 km east of city of Marsa Matruh as shown in Figure 1. The river pours into the Mediterranean Sea and extends approximately 22 km from south to north with varying slope rates [37]. The study area covers approximately 65.02 km^2 and lies between longitudes 27°14′30″ and 27°24′30″ E and latitudes 31°3′30″ and 31°16′0″ N. Wadi El-Garawla has many varieties of environmental conditions typical for that region [38,39].

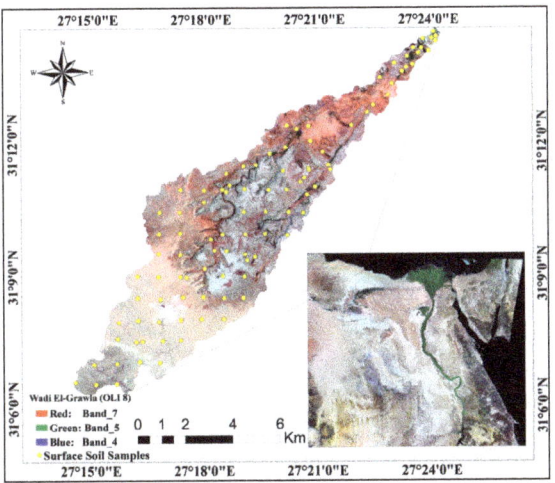

Figure 1. Location of the study area of Wadi El-Garawla and the soil samples as mapped in Landsat 8 satellite imagery (RGB 7, 5, 4).

The rainfall in the studied area ranges between 105.0 to 200 mm/y and the average temperature ranges between 8.1 and 18 °C in the winter and 20 and 29.2 °C in the summer.

The study area is characterized by the scarcity of vegetation cover during the summer and autumn seasons. The vegetation begins to increase at the end of winter and spring, when seasonal herbs and plantings grow depending on the winter precipitation [1,30]. The soil temperature regime of the area is thermic and the soil moisture regime is torrid. In addition, the soils were classified in two orders—Entisols and Aridisols—and divided into five subgroups: Typic Calcigypsids, Typic Haplogypsids, Typic Haplocalcids, Typic TorriPsamments, and Lithic Torriorthents [39].

2.2. Soil Sampling and Chemical Analysis

The soil sample sites were determined based on the characteristics and the heterogeneity of the area because surface properties differ from south to north and were acquired on 15th December 2019. The amount of transported sediments is much deeper in the south. One hundred surface soil samples (0–25 cm) were gathered using a random sampling method. All geomorphic units were represented by several soil samples. The collected samples were dried in the laboratory at a normal temperature and then sifted by a 2 mm sieve. The collected soil samples were chemically analyzed in a laboratory where SOM was analyzed based on Walkley and Black and soil acidity (pH) in soil saturated paste by PH meter according to previous methods [40]. The soil's available N content was measured for each soil sample using conventional chemical analysis via the Kjeldahl method. The available phosphorus content and available potassium content were determined using flame photometry [41].

Table 1 shows the basic statistics of chemical analysis and shows that the soil of the study area is slightly to moderately alkaline with pH values from 6.56 to 8.97. Total soluble salts differed widely from one site to another and had a wide range, as the electrical conductivity of the soil-saturated water (ECe) ranged between 0.11 and 10.53 dS/m. The cation exchange capacity (CEC) also differed from one site to another due to the ratio of the fine fraction and soil organic matter percentage, which ranged between 0.86 and 5.66 cmol/kg. The calcium carbonate percentage of the soils had a wide range, between 2% and 37%. The soil organic matter percentage (SOM%) ranged from almost none (0.04%) to low (1.57%).

Table 1. Basic statistics of chemical analysis of the study area.

	Sand%	Silt%	Clay%	$CaCO_3$%	pH	ECe (dS/m)	CEC (cmol/kg)	SOM%
min	92.14	0.02	2.27	2	6.56	0.11	0.86	0.04
max	96.85	2.91	6.26	37	8.97	10.53	5.66	1.57
mean	94.37	1.31	4.32	19.5	8.01	5.32	2.23	0.38

2.3. Digital Image Processing

Operational land imager (OLI) Landsat 8 images are characterized by 15 m panchromatic and 30 m multi-spectral spatial resolutions with nine spectral bands. Firstly, two OLI images acquired on 15 December 2019 were downloaded from the U.S. Geological Survey (USGS). In particular, the blue to short-wave infrared portion of the spectrum were used in this study. The thermal bands were excluded, and the images were geo-rectified according to UTM coordinates. All further digital image processing and analyses of the OLI satellite images were executed using the standard approaches provided by the ENVI software. Afterwards, all OLI images were atmospherically corrected using the FLAASH module, and the spatial resolution of the visible/NIR bands was resampled to 15 m depending on the panchromatic band. The data were represented by calibration to spectral radiance and then transformed to surface reflectance [41]. The images were mosaicked by combining multiple images into a single composite image within a dereferenced output mosaic. Finally, all satellite images were corrected and matched with the ground measurements of the study.

2.4. Spectral Measurements of the Soil Samples

Analytical spectral devices (ASDs; ASD-4 field spectroradiometer, Boulder, CO, USA) can record a complete range of 350–2500 nm spectrum of 0.1 s. Therefore, an ASD was used to collect spectra over the visible and near-infrared regions for each soil sample at 1.4–2 nm intervals with a spectral resolution of 3–10 nm. The readings were calibrated using the white reference panel. To avoid any change in radiation conditions, the white reference was checked. An ASD spectroradiometer measures the reflectance, transmission, radiance, and irradiance of an object. The recorded data are usually affected by surrounding factors, such as sources of illumination, scanning time, atmospheric conditions, and the field-of-view of the device. Therefore, a contact probe was used to control for those factors in the laboratory. Spectral data were recorded concerning an external white reference panel. Afterwards, five spectra for each sample were recorded, and the average values for the five spectral readings were calculated. Thus, one value was obtained to express the spectral characteristics of each sample [9,42,43].

2.5. Model Calibration and Validation

The spectral modeling of the soil data was achieved using PLSR, which is considered one of the most common approaches in Vis-NIR chemometrics analysis. This method depends on making the relation between the data matrix X and Y through a linear multivariate model [44]. PLSR algorithm integrates the compression and regression steps and selects successive orthogonal factors that maximize the covariance between predictor and response variables [44]. The advantage of PLS regression is that all available wavebands can be incorporated in the model, while earlier studies indicate that PLS models include redundant wavelengths and selecting specific wavebands can refine PLS analyses [45]. The soil samples were representative of the variation soil types in the Wadi El-Garawla basin. Using a leave-one-out cross validation, the dominant absorption features of each soil variable (N, P, K, pH and OM) were determined using PLSR. One hundred soil samples were randomly divided into a subset of 70 samples used for calibration of a subset of 30 samples for validation. Modeling was performed using the PLSR adopted because it usually provides promising results for Vis-NIR analysis [46,47]. The PLSR models (one for each soil parameter) were evaluated by the coefficient of determination (R^2), the root means square error (RMSE), and the mean of response (MR). In addition, R^2 was used to describe the model validation, where "x" represents the soil parameter values (N, P, K, pH, and SOM), which was measured using chemical laboratory analyses and used as the reference values for the calibration phase, "y" is the predicted value, and "n" is the number of soil samples used for the calibration [48,49]. MR, RMSE, and root means square standardized error (RMSSE) were calculated according to Equation (2) [50], and NRMSE was applied according to Equation (3) [51].

$$RMSE = \sqrt{\left(\frac{1}{n}\right)(ps_i - os_i)^2} \quad (1)$$

where n is the total number of samples, and pi is the vector of predicted values of the variable being predicted, with oi being the observed values.

$$RMSSE = \left[\frac{1}{n}\sum_{i=1}^{n}(ps_i - os_i)^2\right]^{\frac{1}{2}} \quad (2)$$

where n is the number of observations or samples; o is the os_i is the standardized observed value at place i; ps_i is the standardized predicted/estimated value at place I

$$NRMSE = \frac{RMSE}{(\delta(y))} \quad (3)$$

where NRMSE is defined as the normalized root mean square error, RMSE as the root mean square error, and $\sigma(y)$ as the standard deviation of y, which is used in [51], where it is explained that the standard deviation (sd)-based NRMSE represents the ratio between the variation not explained by the regression vs. the overall variation in y. Thus, if the regression explains all of the variation in y, nothing is unexplained, and the RMSE, and consequently the NRMSE, is zero. If the regression explains some parts and leaves other parts unexplained, which is at a scale similar to the overall variation, then the ratio will be around 1.

2.6. Mapping Soil Properties Using Ordinary Kriging

Ordinary kriging was used to interpolate the predicted soil values obtained from the PLSR model. OK is a geostatistical model that uses a set of statistical tools to predict the value of a given soil property (N, P, K, pH, and SOM) at a location that was not sampled. The normal distribution pattern of the data was checked using the histogram tool and normal QQPlots. The trend analysis was a check for each parameter.

The general equation of the kriging estimator method is as follows [51,52]:

$$Z^*(x_o) = \sum_{i=1}^{N} \lambda_i Z(X_i) \qquad (4)$$

where $Z^*(x_o)$ is the estimated variable at the x_o location, $Z(x_i)$ represents the values of an inspected variable at the x_i location, λ_i is the statistical weight that is offered to the $Z(x_i)$ sample located near x_o, and N is the number of observations in the neighborhood of the inspected point.

The semivariogram of the selected soil parameter was achieved using the average squared differences among all pairs of values according to Equation (4) [52].

$$\gamma(h) = \frac{1}{2N(h)} \sum_{i=1}^{N(h)} [Z(x_i) - Z(x_i + h)]^2 \qquad (5)$$

where $\gamma(h)$ is the semivariance for the interval distance class h, $N(h)$ is the number of pairs of the lag interval, $Z(x_i)$ is the measured sample value at point i, and $Z(x_i + h)$ is the measured sample value at the position $(i + h)$.

We applied the multiple semi-variogram models (linear plateau, circular, spherical, exponential, exponential, and Gaussian) for each parameter dataset. The validation and suitability of each model was tested via such parameters as the root mean square error (*RMSE*), the mean standardized error (*MSE*), and the root mean square standardized error (*RMSSE*) [50–52]. All data processing and analysis for OK were done in the ArcGIS software package, version 10.4.

2.7. Mapping N, P, K, SOM, and pH Using the Landsat 8 OLI

Spectral reflections obtained by the ASD spectroradiometer were used to determine the N, P, K, SOM, and pH values from the OLI images. The average of the spectral reflectance of the ASD was calculated at regions similar to that of the OLI images: blue (0.450–0.515 µm), green (0.525–0.600 µm), red (0.630–0.680 µm), NIR (0.845–0.885 µm), SWIR1 (1.560–1.660 µm), and SWIR2 (2.100–2.300 µm). PLSR was used for relationship modeling between the averaged values from the ASD spectroradiometer and the N, P, K, pH, and SOM values. Consequently, the models were applied to retrieve N, P, K, pH, and SOM values from OLI images. Thirty soil samples were selected to verify the validity of the results. The randomly selected samples included the spatial diversity of the topographic characteristics of the study area, where the element concentrations differed from one location to another, as they are usually related to the surrounding factors. Finally, the resulting maps were validated by comparing the predicted values with the laboratory values using the correlation coefficient (R^2) and the root mean square error (RMSE). The stepwise linear regression model was used to conduct a regression analysis between the spectral band calculated from the ASD spectroradiometer and soil laboratory analysis in situ for N, P, K, PH, and SOM.

2.8. Soil Fertility Status of Wadi El-Garawla

Soil fertility status (SFS) represents the nutrient content of soil, the available nitrogen, phosphorus, and potassium, the organic matter content, and the soil reaction (pH) (Figure 2) which indicates the degree of suitability for most crops for specific uses. In this study, we relied on the criteria for crop growth and needs, which were suggested in [28,31]. Five factors were selected in this study for evaluating the fertility degree for most crops, as shown in Table 2. The selected factors were the available N, P, and K, the SOM, and the pH, which were produced based on the spectroscopy techniques using the above methodology. Each factor was reclassified using the Arc GIS spatial model, and its weight was taken according to standard methods. The soil fertility status was evaluated using the GIS spatial model based on the following Equation (6) [31].

$$[(S\ ava.\ N \times S\ ava.\ P \times S\ ava\ K \times S\ OM \times S\ pH)]^{(\frac{1}{5})} \qquad (6)$$

where S is the score factor and ava. N, ava P, ava K, OM, and pH are factors that express, respectively.

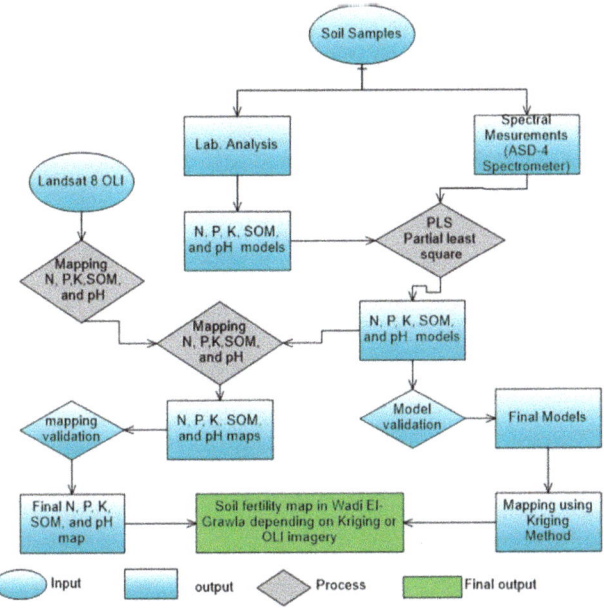

Figure 2. The flowchart methodology of soil nutrient status at Wadi El-Garawla.

Table 2. Criteria of soil fertility and their factor score.

Diagnostic Factor	Unit	1	0.8	0.5	0.2
N	mg/kg	>80	80–40	40–20	<20
P	mg/kg	>15	15–10	10–5	<5
K	mg/kg	>400	400–200	200–100	<100
SOM	g/100 g	>2	1–2	0.5–1	<0.5
pH	-	5.5–7	7–7.8	7.9–8.5	>8.5

Figure 2 shows the steps of mapping different soil nutrients based on the integration of spectral reflectance and satellite images.

3. Results

3.1. Soil Characteristics

Table 3 shows the basic statistical data of the predicted soil analysis (available N, P, and K, as well as the pH and SOM). The maximum values of the N, P, and K, the pH, and the SOM were 60.41, 1.32, 152, 8.97, and 1.05, respectively. On the other hand, the minimum values were 14.04, 0.72, 18, 6.56, and 0.04, and the standard deviations were 8.92, 0.45, 115.6, 0.5, and 0.27, respectively.

Table 3. Statistical parameters of soil properties (N, P, K, pH, and soil organic matter (SOM)).

	Ava. N ppm	Ava. P ppm	Ava. K ppm	pH	SOM%
Min	14.04	0.72	18	6.56	0.04
Max	60.41	2.43	152	8.97	1.57
Mean	37.23	1.58	85	8.01	0.38
Standard deviation	8.92	0.45	115.6	0.5	0.27

3.2. Spectral Characteristics of Studied Soil

The results showed the variance of the spectral reflections of the soil of the study area. Two dominant absorption features were observed in the ultraviolet and near-infrared wavelength ranges (355 and 1080 nm) in response to the N concentration. The change in the curve of the electromagnetic spectrum was associated with changes in the elements and the chemical composition, as the location of the response changed with K, where the response was in the NIR portion of the spectrum at 983 nm. Furthermore, the concentration of phosphorus affects several parts of the spectrum, in the red, SWIR1, and SWIR2 regions (Figure 3).

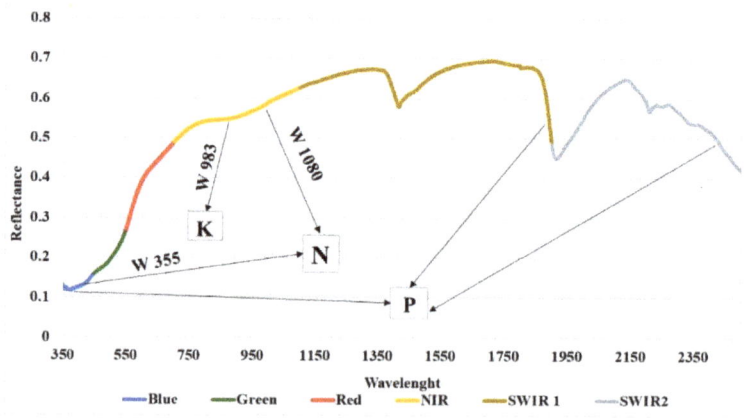

Figure 3. The spectral responses places of soil nutrients (N, P, K), each color refers to the portion of wavelength (blue, green, red, near-infrared (NIR), SWIR1, and SWIR2).

3.3. Prediction of N, P, K, pH, and SOM

The quantitative prediction of the N, P, K, pH, and SOM maps was produced using the PLSR models with an accuracy value (R^2) calibration of 0.89 (N), 0.72 (P), 0.91 (K), 0.65 (pH), and 0.75 (SOM). These models were successfully validated with 30% of the soil samples. The calibration and validation were evaluated by RMSE, MR, NMRSE, and coefficient of determination, as described in Table 4. The validation of the models provided reasonable results for N, P, K, pH, and SOM R^2 validation = 0.87, 0.87, 0.9, 0.69, and 0.84.

Table 4. Statistical parameters of soil properties (N, P, K, pH, and SOM).

Properties	R² Calibration	Adj. R²	RMSE	MR	NRMSE	NRMSE (%)	R² Validation	Spectral Range
Ava. N	0.89	0.86	0.11	1.01	0.01	1.29	0.87	Blue–NIR
Ava. P	0.72	0.7	0.24	1.12	0.69	68.57	0.87	Blue–SWIR1–SWIR2
Ava. K	0.91	0.9	0.24	2.04	0.01	0.56	0.9	NIR
pH	0.65	0.61	0.19	8.02	0.27	26.76	0.69	Blue–Green–SWIR2
SOM	0.75	0.73	0.12	0.41	0.44	44.44	0.84	Blue–SWIR1–SWIR2

Table 5 shows the statistics of the measured and predicted values, including the maximum and minimum values and standard deviation. A variation between the estimated and predicted values was observed, the maximum values of N, P, K, pH, and SOM, and the difference between the measured and predicted values were recognized: they ranged from 60.41 to 56.33 ppm for available N, 1.32–1.81 ppm for available P, 152–151.5 ppm for available K, 8.67–9.79 ppm for pH, and 1.05%–1.23% for SOM. There was also better convergence of the mean values between the measured and predicted values, and the means ranged from 43.1 to 40.2 ppm for available N, 0.96–1.2 ppm for available P, 75.8–76.71 ppm for K, 8.01–8.03 for pH, and 0.39–0.46% for SOM. On the other hand, the minimum values exhibited weak convergence between the measured and predicted values for all factors except pH.

Table 5. Measured and predicted values of N, P, K, pH, and SOM.

	Ava. N (ppm)		Ava. P (ppm)		Ava. K (ppm)		pH		SOM%	
	Measu.	Pred.	Measu.	Pred.	Measu.	Pred.	Measu.	Pred.	Measu.	Pred.
Min	14.04	22.92	0.72	0.44	18	11.39	6.56	7.12	0.04	0.01
Max	60.41	56.33	1.32	1.81	152	151.5	8.97	9.79	1.57	1.23
Mean	43.1	40.2	0.96	1.20	75.8	76.71	8.01	8.03	0.38	0.46
Standard deviation	9.3	8.5	0.16	0.35	41.7	43	0.7	0.71	0.27	0.27

3.4. Mapping of Soil Nutrients of Based on Ordinary Kriging

The mapping of soil nutrient properties was based on retrieving the selected nutrient values based on the spectral fingerprints of each characteristic. The prediction models were applied by previous statistical analysis of N, P, K, pH, and SOM. Thus, OK was used to map the soil nutrients. The performance of ordinary kriging interpolation and the efficiency of the geostatistical model for each soil parameter were checked by such parameters as the RMSE, the MSE, and the RMSSE, as illustrated in Table 6. Results showed that the spherical model was suitable for available N, pH, and SOM, and the Gaussian model was suitable for available P and available K.

Table 6. Geostatistical analyses and semi-varogram parameters.

Soil Properties	Model Type	Mean	Root Mean Square (Rmse)	Mean Standardized (Mse)	Root-Mean-Square Standardized (Rmsse)	Average Standard Error
Ava. N	Spherical	0.133	8.64	0.051	1	8.56
Ava. P	Gaussian	−0.002	0.38	−0.007	0.98	0.38
Ava. K	Gaussian	0.14	27.5	0.004	1.01	27.19
pH	Spherical	0.009	0.44	0.01	0.99	0.44
SOM	Spherical	−0.005	0.22	−0.019	0.97	0.23

Figure 4a–e, respectively, show the spatial distribution of the predicted values of N, P, K, pH, and SOM. The available nitrogen varied from 22.9 to 56.3 ppm. The highest values of nitrogen were located in the north and middle of the study area, where there are agricultural activities. The available phosphorus varied from 0.4 to 2.17 ppm. The available potassium varied from 11.39 to 151.5 ppm. The map of the organic matter showed low SOM content as it varied from 0.01% to 1.23%. The results showed that the predicted pH varied from 7.12 to 9.79 with a mean pH of 8.03. In the current study, the results show that the integration of soil properties gives an acceptable overview of the fertile soil condition distribution.

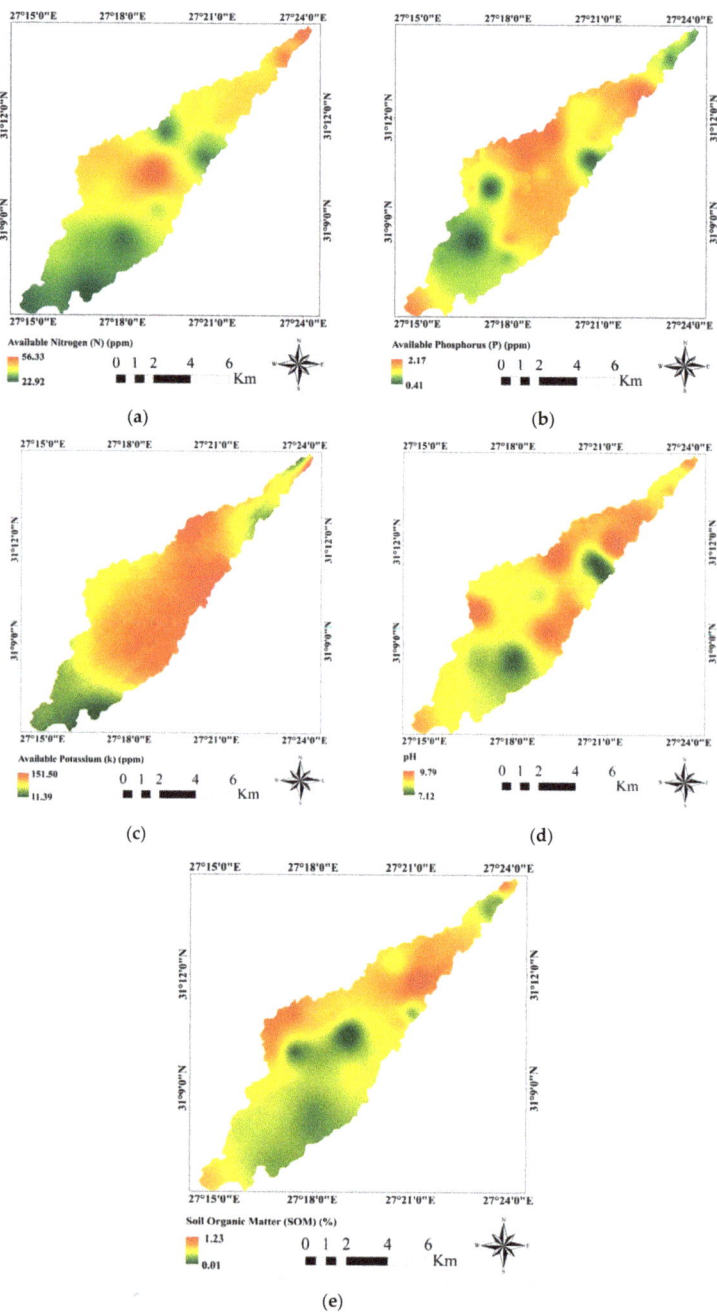

Figure 4. Spatial distribution of predicted soil properties (**a**); predicted available nitrogen (N); (**b**) predicted phosphorus (P); (**c**) predicted available potassium (K); (**d**) predicted pH; (**e**) predicted soil organic matter (SOM) (%).

3.5. Mapping of Soil Nutrients Using Landsat-8 OLI Images

Five equations were obtained, as represented in Equations (6)–(10). The obtained accuracy values for R^2 were 0.923 (N), 0.907 (P), 0.957 (K), 0.978 (pH), and 0.952 (SOM). The five models were applied to OLI imagery to retrieve the spatial distribution values of N, P, K, pH, and SOM. The accuracy assessment was achieved by the NRMSE Table 7.

$$\text{Ava. N} = -31.661 + 186.022 \times \text{Blue} - 364.274 \times \text{Green} + 421.943 \times \text{Red} - 308.068 \times \text{NIR} + 207.957 \times \text{SWIR1} - 12.762 \times \text{SWIR2} \tag{7}$$

$$\text{Ava. P} = 0.404 - 2.702 \times \text{Blue} + 22.540 \times \text{Green} - 14.156 \times \text{Red} + 3.613 \times \text{NIR} - 2.648 \times \text{SWIR1} + 2.304 \times \text{SWIR2} \tag{8}$$

$$\text{Ava. K} = -610.060 - 1424.543 \times \text{Red} + 933.043 \times \text{SWIR2} + 4103.577 \times \text{Green} - 1733.486 \times \text{Blue} \tag{9}$$

$$\text{pH} = 3.983 - 0.544 \times \text{Blue} - 1.112 \times \text{Green} + 6.131 \times \text{Red} + 2.193 \times \text{NIR} - 1.647 \times \text{SWIR1} + 2.739 \times \text{SWIR2} \tag{10}$$

$$\text{SOM} = -1.421 - 8.083 \times \text{Blue} + 17.355 \times \text{Green} - 5.135 \times \text{Red} + 2.473 \times \text{NIR} - 3.275 \times \text{SWIR1} + 2.134 \times \text{SWIR2} \tag{11}$$

Table 7. Model validation of retrieved N, P, K, pH, and SOM from operational land imager (OLI) images.

Properties	RMSE	NRMSE	R^2
Ava. N (ppm)	3.5	0.39	0.71
Ava. P (ppm)	0.06	0.29	0.68
Ava. K (ppm)	4.3	0.076	0.55
pH	0.07	0.22	0.62
Ava. SOM (%)	0.02	0.18	0.7

Each of the variable values (N, P, K, pH, and SOM) were obtained as based on the averages of ASD reflectance spectroscopy and compared to the reflections of the satellite image ranges. Figure 5a–e show the spatial distribution of N, P, K, pH, and SOM, where the available N ranges between 18 and 50 ppm, available P between 0.4 and 2.8 ppm, available K between 9 and 156 ppm, soil PH between 7.30 and 8.28, and SOM between 0.02% and 1.4%. The results of the validation models indicate acceptable outputs for all the elements studied, as the R^2 was 0.7 ± 3.5, 0.68 ± 0.06, 0.55 ± 4.3, 0.62 ± 0.07, and 0.7 ± 0.02 for N, P, K, pH, and SOM, respectively. Meanwhile, the NRMSE values were 0.39, 0.29, 0.076, 0.22, and 0.18 for N, P, K, pH, and SOM respectively, as shown in Table 7.

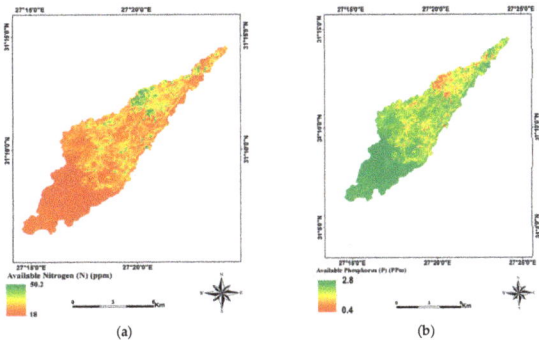

Figure 5. Cont.

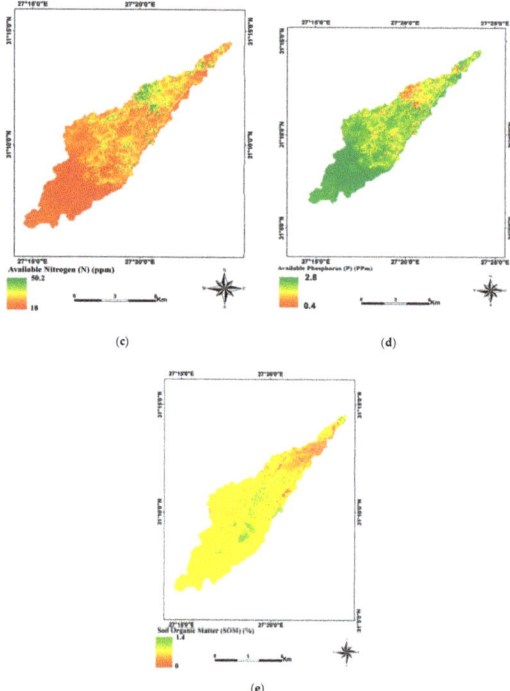

(c)   (d)

(e)

Figure 5. Spatial distribution of predicted soil properties using OLI satellite images: (**a**) nitrogen (N); (**b**) phosphorus (P); (**c**) potassium (K); (**d**) pH; and (**e**) SOM.

3.6. Soil Fertility Status of Wadi El-Garawla

Spatial distribution maps resulting from spectral measurements of soil nutrients were more significant than those produced using satellite imagery. Therefore, fertility status in the study area was evaluated using spectroscopic measurements directly as performed through the spatial modeling of all characteristics. Figure 6 shows the spatial distribution of the status fertility of the study area depending on the integration of available N, P, and K, the SOM, and the pH. Three degrees of soil fertility were recognized in the study area, namely high, moderate, and low. Their respective area of coverage was 1019.136 ha (about 16% of the total study area), 2709.02 ha (43%), and 2536.9 ha (41%).

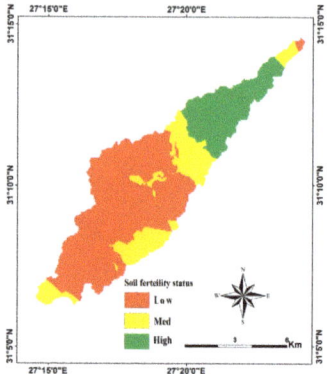

Figure 6. Spatial distribution of soil fertility in Wadi between high, moderate, and low values.

4. Discussion

The PLSR was herein applied to the reflectance spectra measured in the arid conditions of Wadi El-Garawla (on the northwest coast of Egypt) to model, predict, and map the available N, P, and K, pH, and SOM from in situ analysis, Vis-NEARNEAR spectroscopy, and satellite OLI data. Outputs from our analysis indicated that PLSR is reliable and effective for predicting soil nutrients and characteristics, as already found in other diverse areas in the world [16,21,52]. This confirms that the soil's available N, P, and K, the pH, and the SOM can be predicted using Vis-NEAR spectroscopy and shortwave infrared (Vis-NIR-SWIR). The variation in the spectral responses of N, P, and K throughout the wavelength range (350–2500) are associated with the diverse behavior of the diverse elements [18,53].

In addition, the absorption at 1400 and 1900 nm is referred to as the overtone bands related to water and hydroxyls [17,44,54]. On the other hand, the responses to pH were observed in three regions: blue, green, and SWIR2 [55]. Furthermore, the responses for SOM were evident in the blue-SWIR1-SWIR2 regions. As a whole, our findings are consistent with the results of a previous study [56,57]. Spectral reflections are influenced by different soil characteristics and the concentrations of different elements [58]. Even if the determination of the parts of the wavelength that respond to dynamic changes in the concentration of elements is a complicated process, the statistical analysis can overcome these problems.

The obtained models to predict the soil content of nitrogen, phosphorus, and potassium indicate an ability to measure the elements with acceptable accuracy, and these outputs are consistent with many scientists [59–61]. The PLSR model based on the resampled measured spectra as a result of the calibration models can be effectively used to predict N, P, K, pH and SOM values, as is evident by the coefficient of determination: R^2 values were 0.89, 0.72, and 0.91 for N, P, and K and 0.65 and 0.75 for pH and SOM, respectively. The results show the capability of reflectance Vis-NEAR spectroscopy to predict soil pH with a correlation of validation of 0.69, and these findings are consistent with [62–65]. Furthermore, the validation of the models of SOM was 0.48. Hence, the results of the prediction of the SOM content have become acceptable and are consistent with other researchers studying arid and semi-arid areas [66,67]. The high values of pH in the study area refers to an increase in the carbonate contents, and the parent material was limestone. These calcareous lands are common in the western desert in Egypt and the north of Africa [68,69]. Soil pH is a critical factor because it affects the availability of soil nutrients to plant roots, and because it affects the biological activity in different soil environments and the activity of enzymes [70]. The map of available phosphorus shows that the soil of Wadi El-Garawla has low phosphorus content, which may be due to parent lime material. In addition, the available phosphorus is associated inversely with soil pH values, which is consistent with [69–71].

The spatial distribution of SOM in the study area shows that the area, in general, is poor in the proportion of organic matter, less than 0.5% in most of the region. In the northern parts, a relative increase in the percentage of organic matter was observed due to the presence of seasonal crops. Despite the increase of SOM in the north of the area, it was not observed that it had a significant effect on reducing the soil pH in the study area, except for some small areas where a decrease in soil pH with increasing SOM was observed. On the other hand, the results reflect the soil characteristics in Wadi El-Garawla, which strongly vary according to several factors ranging from topographic characteristics, climate conditions, human activities, and soil types [1,72,73]. The accuracy of the spatial distribution maps of soil characteristics using OK show the acceptability degree of the results and their compatibility with other study. The RMSSE values were 1, 0.98, 1.01, 0.99, and 0.97 for N, P, K, pH, and SOM, respectively; this finding agrees with [74]. The RMSSE values were close to one, and the MSE values were close to zero for all parameters. This indicates that OK was appropriate and reliable for predicting the spatial distribution of the studied soil properties.

The study area represents the lands of the northern coast, as it is similar in climatic and topographic conditions and characterized by a low-to-moderate content of available N and P, except in the north where agricultural activities occur, and these results are consistent with those of previous studies [73,75].

The OK map of N, P, and K showed that the predicted values were associated with the SOM content. This large difference in SOM may be due to the variations in topography and climate conditions [76,77].

The integration of Landsat-8 OLI images with spectral measurements provided satisfactory results on the distribution of soil nutrients in Wadi El-Garawla, even though the area is characterized by a low concentration of soil nutrients. These results demonstrate the effectiveness of satellite images (OLI) in predicting different soil properties, and these results are consistent with other independent investigations [18].

The R^2 and root mean square error (Table 6) confirmed the expected results of retrieved soil nutrients by satellite images, where the R^2 for nitrogen and soil organic matter were around 0.71 and 0.7, respectively. Furthermore, the R^2 for phosphorous and soil pH were 0.68 and 0.62, while 0.55 was recorded for available potassium. The accuracy results of OK interpolation were better than the results obtained by satellites, but when using high resolution images satellite imagery, the results may be better than the interpolation method [18]. Wadi El-Garawla is exposed to winter monsoon rains that cause the removal of soil nutrients from the surface layers by the slope effect, where the slope increases from south to north and can cause several environmental hazards, such as drought and desertification [72–80].

However, these low values are due to mismanagement of agricultural activities and the location in a semi-dry climate system. The areas that have significantly high fertility as demonstrated by agricultural activity and natural cover, as reflected in the levels of N, P, and K and organic matter. The results showed that most of the area of Wadi El-Garawla is characterized by low-to-moderately fertile soil, except for some scattered areas to the north that are characterized by high fertility. The south of the area is characterized by shallow-to-very-shallow depths and a coarse texture in addition to an undulated surface topography. The fertile soils are characterized by a deep-to-moderate soil profile, with a flat or almost flat and gently undulating surface to the north [39,78–80].

Agricultural activities on Egypt's northwest coast depend on the availability of water, and that depends on monsoon rains during the winter. The results showed that the Saharan areas in Egypt are poor in their fertile content compared with the soil of Nile Delta [81–84]. This means that the soil of Wadi El-Garawla requires appropriate management that is consistent with the nature of the fertile condition. The types of crops should suit the soil's characteristics and water availability, as well as the climate of the region.

5. Conclusions

VNIR-SWIR spectroscopy is a very helpful technique for evaluating macronutrients, SOM, and soil pH. The current study was based on the modeling of the relationship between the spectral response and the concentrations of different elements. In detail, the PLSR was herein applied to the reflectance spectra measured in the arid conditions of Wadi El-Garawla (on the northwest coast of Egypt) to model, predict, and map the available N, P, and K, pH, and SOM from in situ analysis, Vis-NEAR spectroscopy, and satellite OLI data. Thirty soil samples were selected to verify the validity of the results. The randomly selected samples included the spatial diversity and characteristics of the study area.

As a whole, results from our investigations pointed out that the red and near-infrared regions are the most sensitive portions of the spectrum to N and K concentrations, while the red, SWIR1, and SWIR2 regions are the most sensitive to phosphorus concentrations. On the other hand, the responses to pH occur in three regions: blue, green, and SWIR2. Furthermore, the responses for SOM occur in the blue, SWIR1, and SWIR2 regions. The results indicated that spectroscopy could efficiently predict the concentrations of different elements, with R^2 values of 0.89, 0.72, 0.91, 0.65, and 0.75 for N, P, K, pH, and SOM, respectively. On the other hand, the validation of the models provided reasonable results, where the R^2 and RMSE for N, P, K, pH, and SOM were 0.87 ± 0.11, 0.87 ± 0.24, 0.9 ± 0.24, 0.69 ± 0.19, and 0.84 ± 0.12, respectively. Moreover, the use of Landsat-8 OLI images can produce acceptable results on the spatial distribution of soil nutrients, where R^2 and RMSE were 0.7 ± 3.5, 0.68 ± 0.06, 0.55 ± 4.3, 0.62 ± 0.07, and 0.7 ± 0.02 for N, P, K, pH, and SOM, respectively. Soil fertility in Wadi El-Garawla was classified into three classes: high, moderate, and low, which respectively represented about 16%, 43%, and 41% of the total space of the study area.

The results show that Wadi El-Garawla is characterized by increasing concentrations of soil nutrients in the northern parts due to the removal of these nutrients by active water erosion from the south to the north as a result of the natural slope. The results illustrated the importance of using Vis-NIR for the quantitative prediction of soil nutrients as well as an alternative to chemical analysis procedures. Wadi El-Garawla is characterized by calcareous soils and has a low availability of nutrients. Therefore, the area requires special management that takes into consideration the spatial distribution of nutrients. Suitable crops that can grow in such soils need to be selected. Moreover, it is necessary to rely on organic additives to improve the soil properties, as it helps to facilitate soil nutrients in the calcareous soil.

The methodological approach herein adopted does provide a "fast," low-cost tool that can be promptly applied widely to improve food production efficiency and significantly improve soil conservation and preservation. Our effort is also a contribution to supporting the sustainable management of soil and food production, so it provides a reference for the operational use of Earth Observation (EO)-based tools for addressing ecological problems and poverty alleviation, one of the major global goals of the 2030 Agenda for Sustainable Development (SDGs). As a whole, we proposed that the use of EO for efficient monitoring of the soil conditions can be further improved in the future on a multiscale and multi-parameter scale by integrating EO-based information with socioeconomic factors economy, population, etc., thus offering an integrated system that can be operationally adapted to suitably support and improve the efficiency of local government.

Author Contributions: All the authors substantially contributed to this article. E.S.M., A.A.E.B., A.A.B., A.M.A., and E.S.M conceptualized the study and developed the methodology. The satellite imagery was analyzed by A.A.B., E.S.M, A.M.A., and E.S.M. A.A.E.B., E.S.M., A.E., R.L., and T.E.-b. accomplished data analysis and wrote a draft of the manuscript. A.A.B., E.S.M, A.E., and R.L. contributed to reviewing and editing the manuscript. All authors have read and agreed to the published version of the manuscript.

Funding: The authors would like to declare that the funding of the study has been supported by the National Authority for Remote Sensing and Space Science (NARSS), Egypt and Research Group Project no. RGP-VPP-275., King Saud University, Saudi Arabia.

Acknowledgments: The authors would like to thank the National Authority for Remote Sensing and Space Science (NARSS) for funding the field survey and remote sensing work. The authors would like to thank the Soils and Water Department, Faculty of Agricultural, Tanta University, Egypt, for supervising this work and for sample analysis. The authors would like to thank the Italian National Research Council (CNR) at Potenza and 5-100 Project (RUDN University) and Agrarian-Technological Institute of the Peoples' Friendship University of Russia for supporting the research activities. The authors would like to extend their sincere appreciation to the Deanship of Scientific Research at King Saud University for its funding of this research through the Research Group Project no. RGP-VPP-275.

Conflicts of Interest: The authors would like to hereby certify that there is no conflict of interest in the data collection, the processing of the data, the writing of the manuscript, or the decision to publish the results.

References

1. Mohamed, E.S.; Abu-Hashim, M.; Belal, A.A.A. Sustainable indicators in arid region: Case study–Egypt. In *Sustainability of Agricultural Environment in Egypt*; Springer Cham: Basel, Switzerland, 2018; Part I; pp. 273–293.
2. AbdelRahman, M.A.; Shalaby, A.; Mohamed, E.S. Comparison of two soil quality indices using two methods based on geographic information system. *Egypt. J. Remote Sens. Space Sci.* **2019**, *22*, 127–136. [CrossRef]
3. Abu-hashim, M.; Elsayed, M.; Belal, A.E. Effect of land-use changes and site variables on surface soil organic carbon pool at Mediterranean Region. *J. Afr. Earth Sci.* **2016**, *114*, 78–84. [CrossRef]
4. Hendawy, E.; Belal, A.A.; Mohamed, E.S.; Elfadaly, A.; Murgante, B.; Aldosari, A.A.; Lasaponara, R. The prediction and assessment of the impacts of soil sealing on agricultural land in the North Nile Delta (Egypt) using satellite data and GIS modeling. *Sustainability* **2019**, *11*, 4662. [CrossRef]
5. Mohamed, E.S.; Morgun, E.G.; Bothina, S.G. Assessment of soil salinity in the Eastern Nile Delta (Egypt) using geoinformation techniques. *Mosc. Univ. Soil Sci. Bull.* **2011**, *66*, 11–14. [CrossRef]

6. Hammam, A.A.; Mohamed, E.S. Mapping soil salinity in the East Nile Delta using several methodological approaches of salinity assessment. *Egypt. J. Remote Sens. Space Sci.* **2018**. [CrossRef]
7. Hassan, A.M.; Belal, A.A.; Hassan, M.A.; Farag, F.M.; Mohamed, E.S. Potential of thermal remote sensing techniques in monitoring waterlogged area based on surface soil moisture retrieval. *J. Afr. Earth Sci.* **2019**, *155*, 64–74. [CrossRef]
8. Saleh, A.M.; Belal, A.B.; Mohamed, E.S. Land resources assessment of El-Galaba basin, South Egypt for the potentiality of agriculture expansion using remote sensing and GIS techniques. *Egypt. J. Remote Sens. Space Sci.* **2015**, *18*, S19–S30. [CrossRef]
9. Horta, A.; Malone, B.; Stockmann, U.; Minasny, B.; Bishop, T.F.A.; McBratney, A.B.; Pallasser, R.; Pozza, L. Potential of integrated field spectroscopy and spatial analysis for enhanced assessment of soil contamination: A prospective review. *Geoderma* **2015**, *241*, 180–209. [CrossRef]
10. Angelopoulou, T.; Tziolas, N.; Balafoutis, A.; Zalidis, G.; Bochtis, D. Remote sensing techniques for soil organic carbon estimation: A review. *Remote Sens.* **2019**, *11*, 676. [CrossRef]
11. Chen, F.; Kissel, D.E.; West, L.T.; Adkins, W. Field-scale mapping of surface soil organic carbon using remotely sensed imagery. *Soil Sci. Soc. Am. J.* **2000**, *64*, 746–753. [CrossRef]
12. Rossel, R.V.; Adamchuk, V.I.; Sudduth, K.A.; McKenzie, N.J.; Lobsey, C. Proximal soil sensing: An effective approach for soil measurements in space and time. In *Advances in Agronomy*; Academic Press: Cambridge, MA, USA, 2011; Volume 113, pp. 243–291.
13. Nawar, S.; Buddenbaum, H.; Hill, J.; Kozak, J. Modeling and mapping of soil salinity with reflectance spectroscopy and landsat data using two quantitative methods (PLSR and MARS). *Remote Sens.* **2014**, *6*, 10813–10834. [CrossRef]
14. Abd-Elmabod, S.K.; Alice, F.; Zhang, Z.; Ali, R.R.; Jones, L. Rapid urbanisation threatens fertile agricultural land and soil carbon in the Nile delta. *J. Environ. Manag.* **2019**, *252*, 109668. [CrossRef] [PubMed]
15. Effat, H.A.; El-Zeiny, A.M. Integration of satellite data and spatial decision models for zoning new urban communities in El-Fayoum Desert. *Arab. J. Geosci.* **2020**, *13*, 1–13. [CrossRef]
16. Ben-Dor, E.; Banin, A. Near-infrared analysis as a rapid method to simultaneously evaluate several soil properties. *Soil Sci. Soc. Am. J.* **1995**, *59*, 364–372. [CrossRef]
17. Nawar, S.; Buddenbaum, H.; Hill, J. Digital mapping of soil properties using multivariate statistical analysis and ASTER data in an arid region. *Remote Sens.* **2015**, *7*, 1181–1205. [CrossRef]
18. Nawar, S.; Mouazen, A.M. On-line Vis-NIR spectroscopy prediction of soil organic carbon using machine learning. *Soil Tillage Res.* **2019**, *190*, 120–127. [CrossRef]
19. Mohamed, E.S.; Saleh, A.M.; Belal, A.B.; Gad, A. Application of near-infrared reflectance for quantitative assessment of soil properties. *Egypt. J. Remote Sens. Space Sci.* **2018**, *21*, 1–14. [CrossRef]
20. Khayamim, F.; Wetterlind, J.; Khademi, H.; Robertson, A.J.; Cano, A.F.; Stenberg, B. Using visible and near infrared spectroscopy to estimate carbonates and gypsum in soils in arid and subhumid regions of Isfahan, Iran. *J. Near Infrared Spectrosc.* **2015**, *23*, 155–165. [CrossRef]
21. Abdul Munnaf, M.; Nawar, S.; Mouazen, A.M. Estimation of Secondary Soil Properties by Fusion of Laboratory and On-Line Measured Vis–NIR Spectra. *Remote Sens.* **2019**, *11*, 2819. [CrossRef]
22. Fang, Q.; Hong, H.; Zhao, L.; Kukolich, S.; Yin, K.; Wang, C. Visible and near-infrared reflectance spectroscopy for investigating soil mineralogy: A review. *J. Spectrosc.* **2018**, *2018*, 1–14. [CrossRef]
23. Paz-Kagan, T.; Zaady, E.; Salbach, C.; Schmidt, A.; Lausch, A.; Zacharias, S.; Notesco, G.; Ben-Dor, E.; Karnieli, A. Mapping the spectral soil quality index (SSQI) using airborne imaging spectroscopy. *Remote Sens.* **2015**, *7*, 15748–15781. [CrossRef]
24. El Nahry, A.H.; Mohamed, E.S. Potentiality of land and water resources in African Sahara: A case study of south Egypt. *Environ. Earth Sci.* **2011**, *63*, 1263–1275. [CrossRef]
25. Bayer, A.; Bachmann, M.; Müller, A.; Kaufmann, H. A comparison of feature-based MLR and PLS regression techniques for the prediction of three soil constituents in a degraded South African ecosystem. *Appl. Environ. Soil Sci.* **2012**, *2012*, 1–20. [CrossRef]
26. Islam, K.; Singh, B.; McBratney, A. Simultaneous estimation of several soil properties by ultra-violet, visible, and near-infrared reflectance spectroscopy. *Soil Res.* **2003**, *41*, 1101–1114. [CrossRef]
27. Rossel, R.V.; Walvoort, D.J.J.; McBratney, A.B.; Janik, L.J.; Skjemstad, J.O. Visible, near infrared, mid infrared or combined diffuse reflectance spectroscopy for simultaneous assessment of various soil properties. *Geoderma* **2006**, *131*, 59–75. [CrossRef]

28. Qiao, Y.; Zhang, S. Near-infrared spectroscopy technology for soil nutrients detection based on LS-SVM. In *International Conference on Computer and Computing Technologies in Agriculture*; Springer: Berlin/Heidelberg, Germany, 2011; pp. 325–335.
29. Rossel, R.V.; McBratney, A.B. Laboratory evaluation of a proximal sensing technique for simultaneous measurement of soil clay and water content. *Geoderma* **1998**, *85*, 19–39. [CrossRef]
30. Abd-Elmabod, S.K.; Muñoz-Rojas, M.; Jordán, A.; Anaya-Romero, M.; Phillips, J.D.; Jones, L.; Zhang, Z.; Pereira, P.; Fleskens, L.; Van Der Ploeg, M.; et al. Climate change impacts on agricultural suitability and yield reduction in a Mediterranean region. *Geoderma* **2020**, *374*, 114453. [CrossRef]
31. El Baroudy, A.A. Mapping and evaluating land suitability using a GIS-based model. *Catena* **2016**, *140*, 96–104. [CrossRef]
32. Barbosa, A.; Marinho, T.; Martin, N.; Hovakimyan, N. Multi-Stream CNN for spatial resource allocation: A crop management application. In Proceedings of the IEEE/CVF Conference on Computer Vision and Pattern Recognition Workshops, Seattle, WA, USA, 16–18 June 2020; pp. 58–59.
33. Preston, W.; Araújo do Nascimento, C.W.; Agra Bezerra da Silva, Y.J.; Silva, D.J.; Alves Ferreira, H. Soil fertility changes in vineyards of a semiarid region in Brazil. *J. Soil Sci. Plant Nutr.* **2017**, *17*, 672–685. [CrossRef]
34. Shrivastava, P.; Kumar, R. Soil salinity: A serious environmental issue and plant growth promoting bacteria as one of the tools for its alleviation. *Saudi J. Biol. Sci.* **2015**, *22*, 123–131. [CrossRef]
35. Schütz, L.; Gattinger, A.; Meier, M.; Müller, A.; Boller, T.; Mäder, P.; Mathimaran, N. Improving crop yield and nutrient use efficiency via biofertilization—A global meta-analysis. *Front. Plant Sci.* **2018**, *8*, 2204. [CrossRef] [PubMed]
36. Khdery, G.; Gad, A.A.; El-Zeiny, A.M. Spectroscopic Characterization of Plant Cover in El-Fayoum Governorate, Egypt. *Egypt. J. Soil Sci.* **2020**. [CrossRef]
37. Bellon-Maurel, V.; Fernandez-Ahumada, E.; Palagos, B.; Roger, J.M.; McBratney, A. Critical review of chemometric indicators commonly used for assessing the quality of the prediction of soil attributes by NIR spectroscopy. *TrAC Trends Anal. Chem.* **2010**, *29*, 1073–1081. [CrossRef]
38. Abd-Elmabod, S.K.; Jordán, A.; Fleskens, L.; Phillips, J.D.; Muñoz-Rojas, M.; Van der Ploeg, M.; Anaya-Romero, M.; De la Rosa, D. Modelling agricultural suitability along soil transects under current conditions and improved scenario of soil factors. In *Soil Mapping and Process Modeling for Sustainable Land Use Management*; Elsevier: Amsterdam, The Netherland, 2017; pp. 193–219. ISBN 9780128052006.
39. Khalifa, M.E.; Beshay, N.F. Soil classification and potentiality assessment for some rainfed areas at West of Matrouh, Northwestern Coast of Egypt. *Alex. Sci. Exch. J. Int. Q. J. Sci. Agric. Environ.* **2015**, *36*, 325–341.
40. Soil Survey Staff. Soil survey field and laboratory methods manual. In *Soil Survey Investigations Report No. 51, Version 2.0*; Burt, R., Staff, S.S., Eds.; U.S. Department of Agriculture, Natural Resources Conservation Service: Washington, DC, USA, 2014.
41. Rukun, L. *Analysis Methods of Soil Agricultural Chemistry*; Agriculutural Science and Technology Press: Beijing, China, 1999.
42. Wold, S.; Sjöström, M.; Eriksson, L. PLS-regression: A basic tool of chemometrics. *Chemom. Intell. Lab. Syst.* **2001**, *58*, 109–130. [CrossRef]
43. Chander, G.; Markham, B.L.; Helder, D.L. Summary of current radiometric calibration coefficients for Landsat MSS, TM, ETM+, and EO-1 ALI sensors. *Remote Sens. Environ.* **2009**, *113*, 893–903. [CrossRef]
44. Bilgili, V.; van Es, H.; Akbash, F.; Durak, A.; Hively, W. Visible-Near Infrared Reflectance Spectroscopy for Assessment of Soil Properties in a Semi-Arid Area of Turkey. *J. Arid. Environ.* **2010**, *74*, 229–238. [CrossRef]
45. Kawamura, K.; Tsujimoto, Y.; Rabenarivo, M.; Asai, H.; Andriamananjara, A.; Rakotoson, T. Vis-NIR spectroscopy and PLS regression with waveband selection for estimating the total C and N of paddy soils in Madagascar. *Remote Sens.* **2017**, *9*, 1081. [CrossRef]
46. Helder, D.; Markham, B.; Morfitt, R.; Storey, J.; Barsi, J.; Gascon, F.; Clerc, S.; LaFrance, B.; Masek, J.; Roy, D.P.; et al. Observations and Recommendations for the Calibration of Landsat 8 OLI and Sentinel 2 MSI for improved data interoperability. *Remote Sens.* **2018**, *10*, 1340. [CrossRef]
47. Geladi, P.; Kowalski, B.R. Partial least-squares regression: A tutorial. *Anal. Chim. Acta* **1986**, *185*, 1–17. [CrossRef]
48. Dangal, S.R.; Sanderman, J.; Wills, S.; Ramirez-Lopez, L. Accurate and Precise Prediction of Soil Properties from a Large Mid-Infrared Spectral Library. *Soil Syst.* **2019**, *3*, 11. [CrossRef]

49. Mohamed, E.; Belal, A.A.; Ali, R.R.; Saleh, A.; Hendawy, E.A. Land degradation. In *The Soils of Egypt*; Springer: Basel, Switzerland, 2019; pp. 159–174.
50. Yang, Y.; Zhu, J.; Tong, X.; Wang, D. The spatial pattern characteristics of soil nutrients at the field scale. In *International Conference on Computer and Computing Technologies in Agriculture*; Springer: Boston, MA, USA, 2008; pp. 125–134.
51. Otto, S.A.; Kadin, M.; Casini, M.; Torres, M.A.; Blenckner, T. A quantitative framework for selecting and validating food web indicators. *Ecol. Indic.* **2018**, *84*, 619–631. [CrossRef]
52. Biswas, A.; Si, B.C. Model averaging for semivariogram model parameters. *Adv. Agrophys. Res.* **2013**, *4*, 81–96.
53. El-Alfy, M.A.; Hasballah, A.F.; El-Hamid, H.T.A.; El-Zeiny, A.M. Toxicity assessment of heavy metals and organochlorine pesticides in freshwater and marine environments, Rosetta area, Egypt using multiple approaches. *Sustain. Environ. Res.* **2019**, *29*, 19. [CrossRef]
54. Mohamed, E.S.; Belal, A.; Shalaby, A. Impacts of soil sealing on potential agriculture in Egypt using remote sensing and GIS techniques. *Eurasian Soil Sci.* **2015**, *48*, 1159–1169. [CrossRef]
55. He, Y.; Huang, M.; García, A.; Hernández, A.; Song, H. Prediction of soil macronutrients content using near-infrared spectroscopy. *Comput. Electron. Agric.* **2007**, *58*, 144–153. [CrossRef]
56. Mohamed, E.S.; Ali, A.M.; El Shirbeny, M.A.; El Razek, A.A.A.; Savin, I.Y. Near infrared spectroscopy techniques for soil contamination assessment in the Nile Delta. *Eurasian Soil Sci.* **2016**, *49*, 632–639. [CrossRef]
57. Tekin, Y.; Tumsavas, Z.; Mouazen, A.M. Effect of moisture content on prediction of organic carbon and pH using visible and near-infrared spectroscopy. *Soil Sci. Soc. Am. J.* **2012**, *76*, 188–198. [CrossRef]
58. Zhang, Z.; Yu, D.; Shi, X.; Warner, E.; Ren, H.; Sun, W.; Tan, M.; Wang, H. Application of categorical information in the spatial prediction of soil organic carbon in the red soil area of China. *Soil Sci. Plant Nutr.* **2010**, *56*, 307–318. [CrossRef]
59. Mohamed, E.S.; Ali, A.; El-Shirbeny, M.; Abutaleb, K.; Shaddad, S.M. Mapping soil moisture and their correlation with crop pattern using remotely sensed data in arid region. *Egypt. J. Remote Sens. Space Sci.* **2019**. [CrossRef]
60. Zhai, Y.; Cui, L.; Zhou, X.; Gao, Y.; Fei, T.; Gao, W. Estimation of nitrogen, phosphorus, and potassium contents in the leaves of different plants using laboratory-based visible and near-infrared reflectance spectroscopy: Comparison of partial least-square regression and support vector machine regression methods. *Int. J. Remote Sens.* **2012**, *34*, 2502–2518.
61. Mohamed, E.S.; Schütt, B.; Belal, A. Assessment of environmental hazards in the north western coast-Egypt using RS and GIS. *Egypt. J. Remote Sens. Space Sci.* **2013**, *16*, 219–229. [CrossRef]
62. Stenberg, B.; Rossel, R.A.V.; Mouazen, A.M.; Wetterlind, J. Visible and near infrared spectroscopy in soil science. In *Advances in Agronomy*; Academic Press: Cambridge, MA, USA, 2010; Volume 107, pp. 163–215.
63. Recena, R.; Fernández-Cabanás, V.M.; Delgado, A. Soil fertility assessment by Vis-NIR spectroscopy: Predicting soil functioning rather than availability indices. *Geoderma* **2019**, *337*, 368–374. [CrossRef]
64. Dick, W.A.; Cheng, L.; Wang, P. Soil acid and alkaline phosphatase activity as pH adjustment indicators. *Soil Biol. Biochem.* **2000**, *32*, 1915–1919. [CrossRef]
65. Tumsavas, Z. Possibility of determining soil pH using visible and near-infrared (Vis-NIR) spectrophotometry. *Environ. Biol.* **2017**, *38*, 1095–1100. [CrossRef]
66. Vohland, M.; Ludwig, M.; Harbich, M.; Emmerling, C.; Thiele-Bruhn, S. Using variable selection and wavelets to exploit the full potential of visible–near infrared spectra for predicting soil properties. *J. Near Infrared Spectrosc.* **2016**, *24*, 255–269. [CrossRef]
67. Al-Shujairy, Q.T.A.; Ali, N.S. Prediction of soil available phosphorous content using spectra-radiometer and gis in southern of Iraq. *Iraqi J. Agric. Sci.* **2017**, *48*, 171–177.
68. Reeves, J.B.; McCarty, G.W. Quantitative analysis of agricultural soils using near infrared reflectance spectroscopy and a fibre-optic probe. *J. Near Infrared Spectrosc.* **2001**, *9*, 25–34. [CrossRef]
69. El-Aziz, S.H.A. Evaluation of land suitability for main irrigated crops in the North-Western Region of Libya. *Eurasian J. Soil Sci.* **2018**, *7*, 73–86. [CrossRef]
70. Ostovari, Y.; Ghorbani-Dashtaki, S.; Bahrami, H.A.; Abbasi, M.; Dematte, J.A.M.; Arthur, E.; Panagos, P. Towards prediction of soil erodibility, SOM and CaCO3 using laboratory Vis-NIR spectra: A case study in a semi-arid region of Iran. *Geoderma* **2018**, *314*, 102–112. [CrossRef]

71. Yu, X.; Liu, Q.; Wang, Y.; Liu, X.; Liu, X. Evaluation of MLSR and PLSR for estimating soil element contents using visible/near-infrared spectroscopy in apple orchards on the Jiaodong peninsula. *Catena* **2016**, *137*, 340–349. [CrossRef]
72. AbdelRahman, M.A.; Natarajan, A.; Hegde, R.; Prakash, S.S. Assessment of land degradation using comprehensive geostatistical approach and remote sensing data in GIS-model builder. *Egypt. J. Remote Sens. Space Sci.* **2019**, *22*, 323–334. [CrossRef]
73. Belal, A.A.; Mohamed, E.S.; Abu-hashim, M.S.D. Land Evaluation Based on GIS-Spatial Multi-Criteria Evaluation (SMCE) for Agricultural Development in Dry Wadi, Eastern Desert, Egypt. *Int. J. Soil Sci.* **2015**, *10*, 100–116. [CrossRef]
74. Aldabaa, A.; Yousif, I.A.H. Geostatistical approach for land suitability assessment of some desert soils. *Egypt. J. Soil Sci.* **2020**, *60*, 195–205. [CrossRef]
75. Zaurov, D.E.; Perdomo, P.; Raskin, I. Optimizing soil fertility and pH to maximize cadmium removed by Indian mustard from contaminated soils. *J. Plant Nutr.* **1999**, *22*, 977–986. [CrossRef]
76. Hopkins, B.; Ellsworth, J. Phosphorus availability with alkaline/calcareous soil. In Proceedings of the Western Nutrient Management Conference, Reno, NV, USA, 7–8 March 2005; Volume 6, pp. 88–93.
77. Clark, R.N.; King, T.V.; Klejwa, M.; Swayze, G.A.; Vergo, N. High spectral resolution reflectance spectroscopy of minerals. *J. Geophys. Res. Space Phys.* **1990**, *95*, 12653–12680. [CrossRef]
78. Abdel-Kader, F.H. Digital soil mapping at pilot sites in the northwest coast of Egypt: A multinomial logistic regression approach. *Egypt. J. Remote Sens. Space Sci.* **2011**, *14*, 29–40. [CrossRef]
79. Mirzaee, S.; Ghorbani-Dashtaki, S.; Mohammadi, J.; Asadi, H.; Asadzadeh, F. Spatial variability of soil organic matter using remote sensing data. *Catena* **2016**, *145*, 118–127. [CrossRef]
80. Said, M.E.S.; Ali, A.; Borin, M.; Abd-Elmabod, S.K.; Aldosari, A.A.; Khalil, M.; Abdel-Fattah, M.K. On the Use of Multivariate Analysis and Land Evaluation for Potential Agricultural Development of the Northwestern Coast of Egypt. *Agronomy* **2020**, *10*, 1318. [CrossRef]
81. Barseem, M.; El Tamamy, A.; Masoud, M. Hydrogeophyical evaluation of water occurrences in El Negila area, Northwestern coastal zone–Egypt. *J. Appl. Sci. Res.* **2013**, *9*, 3244–3262.
82. Abdel-Fattah, M.K.; Abd-Elmabod, S.K.; Aldosari, A.A.; Elrys, A.S.; Mohamed, E.S. Multivariate Analysis for Assessing Irrigation Water Quality: A Case Study of the Bahr Mouise Canal, Eastern Nile Delta. *Water* **2020**, *12*, 2537. [CrossRef]
83. Mohamed, E.S. Spatial assessment of desertification in north Sinai using modified MEDLAUS model. *Arab. J. Geosci.* **2013**, *6*, 4647–4659. [CrossRef]
84. Mohamed, E.S.; Abdellatif, M.A.; Abd-Elmabod, S.K.; Khalil, M.M. Estimation of surface runoff using NRCS curve number in some areas in northwest coast, Egypt. In Proceedings of the E3S Web of Conferences, Barcelona, Spain, 10–12 February 2020; EDP Sciences. Volume 167, p. 02002.

Publisher's Note: MDPI stays neutral with regard to jurisdictional claims in published maps and institutional affiliations.

© 2020 by the authors. Licensee MDPI, Basel, Switzerland. This article is an open access article distributed under the terms and conditions of the Creative Commons Attribution (CC BY) license (http://creativecommons.org/licenses/by/4.0/).

Article

Investigating Detection of Floating Plastic Litter from Space Using Sentinel-2 Imagery

Kyriacos Themistocleous [1,2,*], Christiana Papoutsa [1,2], Silas Michaelides [1,2] and Diofantos Hadjimitsis [1,2]

1. ERATOSTHENES Centre of Excellence, Saripolou 2-6, Achilleos 2 Building, Lemesos 3036, Cyprus; christiana.papoutsa@cut.ac.cy (C.P.); silas.michaelides@cut.ac.cy (S.M.); d.hadjimitsis@cut.ac.cy (D.H.)
2. Department of Civil Engineering and Geomatics, Cyprus University of Technology, 30 Arch. Kyprianos Str., Lemesos 3036, Cyprus
* Correspondence: k.themistocleous@cut.ac.cy; Tel.: +357-99570178

Received: 10 July 2020; Accepted: 14 August 2020; Published: 17 August 2020

Abstract: Plastic litter floating in the ocean is a significant problem on a global scale. This study examines whether Sentinel-2 satellite images can be used to identify plastic litter on the sea surface for monitoring, collection and disposal. A pilot study was conducted to determine if plastic targets on the sea surface can be detected using remote sensing techniques with Sentinel-2 data. A target made up of plastic water bottles with a surface measuring 3 m × 10 m was created, which was subsequently placed in the sea near the Old Port in Limassol, Cyprus. An unmanned aerial vehicle (UAV) was used to acquire multispectral aerial images of the area of interest during the same time as the Sentinel-2 satellite overpass. Spectral signatures of the water and the plastic litter after it was placed in the water were taken with an SVC HR1024 spectroradiometer. The study found that the plastic litter target was easiest to detect in the NIR wavelengths. Seven established indices for satellite image processing were examined to determine whether they can identify plastic litter in the water. Further, the authors examined two new indices, the Plastics Index (PI) and the Reversed Normalized Difference Vegetation Index (RNDVI) to be used in the processing of the satellite image. The newly developed Plastic Index (PI) was able to identify plastic objects floating on the water surface and was the most effective index in identifying the plastic litter target in the sea.

Keywords: Sentinel-2; satellite images; plastic litter; spectral indices; spectroscopy; remote sensing; UAVs

---

1. Introduction

Marine litter refers to waste originating from human activities that has been discharged into coastal or marine environments. Such litter may result from activities on either land or at sea [1]. Currently, 60 to 80% of such marine litter consists of plastic, reaching 95% in some areas and has become a serious environmental hazard [2–18]. Based on its weight and shape, marine litter can be classified as floating litter and sinking litter [13]. It has been estimated that marine litter is split into 15% floating on the sea surface, another 15% remains in the water column and 70% subsides on the sea floor [19].

Although there is limited data on plastic inputs in the oceans [20], it is estimated that almost 8 million tons of plastic enter the oceanic ecosystem every year [21]. Other approximations estimate that the oceans may already contain more than 150 million tons of plastic [22]; around 250,000 tons of these contaminants is fragmented into 5 trillion plastic pieces, which may be floating on the oceans' surface [23]. It has also been calculated that every year, between 4.8 and 12.7 million tons of plastic find their way into the ocean from coastal populations worldwide [16], while the Ellen Macarthur Foundation [24] estimates that approximately 25 million tons of plastic end up in the ocean. What is more alarming is the projection that the global quantity of plastic in the ocean will nearly double to 250 million tons by 2025 [16]. It is expected that by 2050, the ocean will contain more plastic by weight than fish [24].

Since plastics have low density, they float on the surface of water bodies, often accumulating into clusters which can be transported over long distances by the prevailing winds and oceanic currents before sinking [25–34]. A large portion of these plastic clusters enter ocean gyres and can result in clusters that are up to several kilometers in size, such as the Great Pacific Garbage Patch (GPGP) [35,36]. Plastic debris is mostly found in coastal areas, especially in front of river mouths and coastal cities [37]. Plastics break down into debris through a combination of several processes, among which are mechanical weathering, biodegradation and photo- and thermal-oxidative degradation by ultraviolet (UV) radiation. It is worth noting that complete mineralization of such plastic debris may not be possible or may take place after hundreds or thousands of years [38–41]. Plastic litter is harmful to marine life, as it results to deformation, suffocation and death [10,42–44] as well as allowing the spread of invasive species and the release of toxic chemicals into the environment [45–47]. Plastic litter tends to be more harmful near coastlines, where biological diversity and species abundance tend to be highest [48–50].

In order to address the issue of plastic marine litter, several programs and directives have been instituted. The European Union (EU) has issued several directives that are related to reducing plastic litter, primarily through the EU Marine Strategy Framework Directive (Directive 2008/56/EC) and the EU Water Framework Directive (Directive 2000/60/EC). The United Nations Environment Programme (UNEP): Regional Seas Programme is an action-oriented agenda that implements region-specific activities, bringing together stakeholders including governments, scientific communities and civil societies [51]. The mandate of the UNEP is to coordinate 18 Regional Seas Conventions and Action Plans, in which 146 countries participate. The UNEP Regional Seas Programme strives to maintain, restore and enhance marine and coastal resources to support human well-being through sustainable development [51]. The United Nations 2030 Agenda for Sustainable Development addresses the issue of plastic litter in water bodies through Sustainable Development Goals such as Goal 6 on "Clean Water and Sanitation" and Goal 12 on "Sustainable Consumption and Production"; these will also contribute to addressing the issue of marine plastic pollution, as the global nature of plastic supply chains dictates a cooperation between nations and across regions [52]. Therefore, it is necessary to establish harmonized definitions and share data and research on marine plastics and microplastics [52,53].

Although marine litter is a worldwide problem, it has not been adequately addressed in the Mediterranean area [54]. A high concentration of plastics is found within the Mediterranean Sea, with the highest amounts of municipal solid waste generated annually per person of 208–760 kg/year [55–62]. The Mediterranean Sea also ranks fourth in the list of oceans with the highest concentration of floating marine litter in the world, with 22,000 tons [62,63]. This is due to the interaction of a number of factors, including that the Mediterranean is essentially a closed basin with limited exchange of water with other oceanic bodies (this is primarily accomplished through the Straits of Gibraltar), combined with inadequate environmentally sound urban waste management systems, considerable marine vessel transportation in coastal waters, negligible tidal flow and heavily populated coastal areas [21,63,64]. According to Pasternak et al. [60], sites along the shores of the Mediterranean show the greatest densities of marine debris in the world. In particular, the Levantine sub-basin, in which Cyprus is situated, has very little interaction with the rest of the Mediterranean [65]. Plastics that enters the sub-basin from surrounding countries (Cyprus, Egypt, Israel, Lebanon, Syria and Turkey) are also washed up on the beaches of these countries [66,67].

Open-water and shoreline surveys that are designed to assess the distribution of plastic debris in oceans and lakes are time-consuming and costly; in addition, they provide only limited aerial coverage and temporal resolution [68]. Although research on how remote sensing can be used to monitor plastic debris in the sea is still in its early stages [29,67,69], several research studies have used various remote sensing methods to identify plastics in the sea [29,68–73]. Satellite images can be used to identify plastics in the water, such as Sentinel-1A and COSMO-Sky-Med Sar images [74], C-Band Radarsat-1 SAR images [75] as well as Landsat TM and EMT+ satellite images [76–78].

Several studies have been conducted to examine marine litter in the Mediterranean. Mansui et al. [63] simulated marine litter drift in the Mediterranean and noted permanent accumulations of plastics; their study found that the coastline between Tunisia and Syria had the highest amount of plastic accumulation, while the western Mediterranean demonstrated a rather low coastal impact [79]. The University of the Aegean [80] conducted a study to detect and track plastic targets on the sea surface using UAV and images from Sentinel satellites. Their study used a 'target' that was 100 m^2 composed of 1.5 L water bottles, plastic bags and nylon fishing nets. The objective of the study by the University of the Aegean was to evaluate the ability of satellites to detect marine litter on the sea surface using image analysis, image-processing algorithms and satellite measurements [80]. Research has found that large plastic debris can be identified using unmanned aerial vehicles (UAVs) [81–83]. More recently, research has focused on the ability to detect plastic using spectroscopy and high spatial resolution multispectral imaging [84]. Research indicates that near to shortwave infrared (NIR–SWIR) imaging from UAV platforms can be used to detect plastic in the water [79,80].

The objective in this study is to examine whether Sentinel-2 satellite images were effective in identifying plastic clusters in the sea. As well, various indices used for satellite image processing were examined in order to determine whether they were able to identify plastic litter in the water. Spectroscopy was used in order to acquire and compare the spectral signature of the water and the plastic litter.

2. Materials and Methods

In this study, plastic bottles were employed in order to determine if plastic litter can be identified through Sentinel-2 satellite images. According to Biermann et al [29], the high spatial resolution for up to 10 m × 10 m is able to detect small features in the sea, such as plastic litter. In this study, the objective was to examine if a smaller target can be detected by the Sentinel-2 satellite images. Aerial and Sentinel-2 satellite images were used to identify plastic litter in the water. A UAV platform with multispectral cameras took photos of the plastic litter target at the same time as the Sentinel-2 satellite overpass in order to examine different wavelengths in which plastic could be detected in seawater.

The study was conducted in Limassol, Cyprus, south of the Limassol Old Port. The site was selected as this area accumulated a large amount of debris from the ships waiting to enter Limassol Port. Research indicates that plastic debris is often found in areas with high marine traffic [13,53,70,71]. Plastic bottles comprise most of the floating marine debris and accumulate on the bottom of the sea and wash up on the coastlines [62,84]. The International Coastal Cleanup (ICC) report [85] found that plastic bottles were the third most common type of beach litter, that 10% of the global marine debris is plastic bottles [86] and make up 14% of the Mediterranean debris [55]. In order to identify plastic litter in the water, a plastic litter "target" measuring 3 m × 10 m was created from water bottles of 0.5-liter and 1.5-liter size (Figure 1), to emulate marine plastic litter clusters floating in the sea. Plastic bottles were utilized as they are considered to comprise one of the most common types of marine debris [62,85]. Pasternak's study [60] on plastic bottles found that large bottles were 6.3 times more abundant than bottles smaller than 1.5 liter [87]. The plastic litter target used for this project consisted of 1500 plastic bottles that were held together by nylon string and framed by PVC pipes.

The study took place on 15 December 2018 at the Limassol Old Port, during the Sentinel-2 satellite overpass. The use of Sentinel-2 images was selected as the images are free and open to all users through the Copernicus Hub. The satellite acquires images every 5 days, so that the time series of satellite images can be easily acquired for operational time-series applications. During the Sentinel-2 satellite overpass, the plastic litter 'target' was placed into the sea near the Old Port in Limassol, Cyprus and moved 200 m from the coastline (Figure 2).

Figure 1. Plastic litter 'target' made from water bottles. Kyriacos Themistocleous.

Figure 2. Plastic litter "target" being lowered into the sea at Limassol Old Port. Kyriacos Themistocleous.

Divers moved the target to 200 m south of the Limassol shoreline over a depth of 20 m, in order to simulate plastic marine debris in the ocean (Figure 3). A GPS tracker was attached to the plastic litter target to monitor the location of the target during the Sentinel-2 MSIL1C satellite overpass, which occurred on 15 December 2018 at 08:58 UTC.

The SVC HR-1024 spectroradiometer was used to obtain the spectral signatures of the water surface and the plastic litter "target" after it was placed in the water by taking measurements from the top of the wharf. The spectral signatures were taken from 1.5 and 3 m height to check the spectral response at different heights. The spectral signatures were measured with 4 degrees field of view with a sample area of 10 cm and 20 cm. Approximately 20 sampling points at each height were taken at increments of 1 meter apart perpendicular to the target. The multiple spectra taken at each sampling point were averaged by height in spectral groups to account for the effect of wind, water movement, etc.

Figure 3. Divers moving target 200 meters from shoreline. Kyriacos Themistocleous.

During the Sentinel-2 satellite overpass, two UAVs were flown over the plastic litter target to acquire aerial images of the target at a low altitude. The DJI Phantom 4 Pro with an integrated RGB (red, green, blue) 20MP camera (Figure 4, left) and the DJI Phantom 2 with a modified GoPro camera and a Sony Exmor IMX206 multispectral camera with 660 nm and 850 nm filters (Figure 4, right) were utilized. The aerial images from the UAVs were compared to the images from the Sentinel-2 satellite. A significant limitation for the detection of plastics in water through the adoption of NIR spectroscopy and multispectral sensors is the strong absorption of infrared radiation by water [88]. The plastic bottles were collected and disposed in a plastic recycling bin at the end of the study.

Figure 4. Left: Phantom 4 Pro with RGB integrated 20MP camera. Kyriacos Themistocleous.
Right: Phantom 2 with Sony Exmor and modified GoPro camera. Kyriacos Themistocleous.

The Sentinel-2 satellite is able to generate multispectral data with 13 bands in the visible, near infrared and short-wave infrared part of the spectrum with a spatial resolution of 10 m, 20 m and 60 m, as shown in Table 1. It can detect patches of plastic litter of various sizes [29]. Several studies use SWIR imagery to detect plastics in the sea [70]; however, the spatial resolution of 60 m × 60 m would be too low in order to identify a target of 3 m × 10 m. The satellite overpass occurs every 5 days and provides a systematic coverage of the entire Mediterranean Sea. The Sentinel-2 satellite was employed because (a) it provides free and open data, (b) it provides a systematic overview at the same location since the overpass occurs every five days, which provides the ability to revisit a specific location, and (c), the pixel size of the Sentinel-2 satellite has better spatial resolution from other satellite with freely available data, especially in the visible and NIR bands. In this study, the images generated from the multispectral camera were compared with the corresponding wavelengths of the Sentinel-2 bands. Atmospheric correction was applied using the Sen2Cor processor but without any results for identifying the plastic litter target since plastic bottles are transparent and, by applying atmospheric correction, the values of water and floating plastic bottles are absorbed by the atmospheric correction algorithm; therefore, atmospheric correction was not applied. The 2 MSIL1C satellite was selected as atmospheric correction was not required.

Table 1. Wavelengths of Sentinel-2 bands.

Band Name	Spatial Resolution (m)	Central Wavelength (nm)	Bandwidth (nm)
B01	60	443	20
B02	10	490	65
B03	10	560	35
B04	10	665	30
B05	20	705	15
B06	20	740	15
B07	20	783	20
B08	10	842	115
B08A	20	865	21
B9	60	945	20
B10	60	1375	30
B11	20	1610	90
B12	20	2190	180

Research by Rokni et al. [89] proposed several well-established indices for water features extraction. Therefore, equations 1–6 were applied to examine if plastics can be detected in water. The Simple Ratio equation employed blue and NIR bands to examine the bands that were identified in the field measurements from the spectral signatures. The Normalized Difference Water Index (NDWI) [90], Water Ratio Index (WRI) [91], Normalized Difference Vegetation Index (NDVI) [92], Automated Water Extraction Index (AWEI) [93], Modified Normalization Difference Water Index (MNDWI) [94] and Normalization Difference Moisture Index (NDMI) [95], as indicated in Equations (1)–(6) and the Simple Ratio (SR) was also used, delineated by Equation (7). In order to justify the potential of detection of the plastic target by introducing purpose-built relationships, the Plastic Index (PI) and Reversed Normalized Difference Vegetation Index (RNDVI), as expressed by Equations (8) and (9), respectively, were developed in this research effort in order to examine the use of the specific wavelength identified from the spectral signatures.

$$NDWI = (B03 - B08)/(B03 + B08) \tag{1}$$

$$WRI = (B03 + B04)/(B08 + B012) \tag{2}$$

$$NDVI = (B08 - B04)/(B08 + B04) \tag{3}$$

$$AWEI = 4 \times (B03 - B012) - (0.25 \times B08 + 2.75 \times B011) \tag{4}$$

$$MNDWI = (B03 - B012)/(B04 + B012) \tag{5}$$

$$NDMI = (B03 - B08)/(B03 + B08) \tag{6}$$

$$SR = B08/B04 \tag{7}$$

$$PI = B08/(B08 + B04) \tag{8}$$

$$RNDVI = (B04 - B08)/(B04 + B08) \tag{9}$$

Due to the innovative nature of the study, a sensitivity analysis was used to better understand the dynamics and emergent patterns of the indices and provide derivatives of model output parameters (observables) with respect to input parameters [96]. The sensitivity analysis was developed according to the parameters and carried out on the above indices to estimate the objective sensitivity measures in the form of partial derivatives of the model outcomes with respect to input parameters. Each index was examined based on the minimum and maximum values within the Area of Interest, the number of pixels detected for the plastic litter target and the discriminative value, which is the distinct separation that results from the maximum water values and the minimum plastic values for each index. These values were normalized by dividing the maximum value subtracted from the minimum value in the Area of Interest in order to determine the sensitivity parameters of each index. The Area of Interest was selected around the target with a buffer of 50 m radius, which was approximately an area of 8000 m^2, as shown in Figure 9.

3. Results

After the plastic litter target was placed in the water and its location was fixed using anchors, the UAVs were flown over the target in order to acquire RGB and infrared (B08) images of the target at the same time as the Sentinel-2 satellite overpass. Both UAVs were flown over the target so that the images would be taken at the same time. Figure 5 features aerial images of the plastic litter target in both RGB 20MP and Infrared (B08). The RGB image of the plastic litter target (Figure 5, left) is in natural color while the infrared image of the plastic litter target (Figure 5, right) is in black and white for visual comparison.

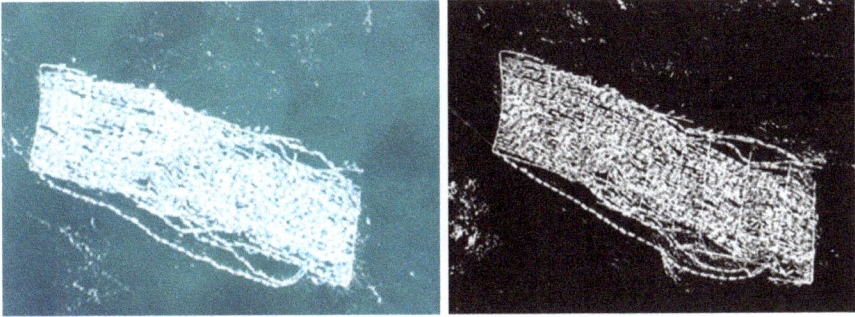

Figure 5. Left: RGB image of plastic litter target. **Right**: infrared image of plastic litter target.

The multispectral image at 660 nm (Figure 6, left) indicates that the reflectance of the water and the reflectance of the plastic litter target have lesser variance than the multispectral image at 850 nm (Figure 6, right) which indicates the reflectance of the plastic litter target, which results from the water's increased absorption of solar radiation.

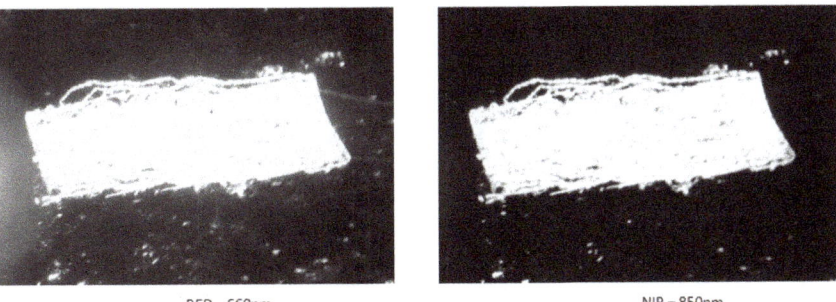

Figure 6. Left: target as evident at 660 nm. **Right**: target as evident at 850 nm (Sony Exmor camera).

Spectral signatures show high reflectance in plastics and no reflectance for water in the near-infrared (NIR) domain [29,85,97]. NIR spectroscopy is currently used in related applications, including the sorting of plastic debris in recycling facilities [66,67,78,84,98]. After placing the plastic litter target in the water, the spectral signatures of the sea water and the plastic bottles were taken and plotted according to the different channels of the Sentinel-2 satellite. The results of the study indicated that plastic bottles have high reflectance in the blue and NIR bands and the water has high reflectance in the blue and low reflectance in the NIR bands. Although, theoretically, the reflectance of water should be zero in the infrared wavelength, due to sediment and debris, there is a low reflectance for water at those wavelengths. In this study, the plastic litter detection was only examined between 400–900 nm. Within this range, the reflectance of the plastic bottles is low in the red band, but increases in the infrared bands, where water absorbs all solar radiation and has almost no reflectance [68], as indicated in Figure 7. Therefore, plastic bottles can be identified using B06, B07 and B08 of Sentinel-2 satellite images. The 20 m spatial resolution in bands B06 and B07 makes it difficult to detect smaller targets, while B08 has a 10 m spatial resolution, which is able to detect the objects due to higher spatial resolution.

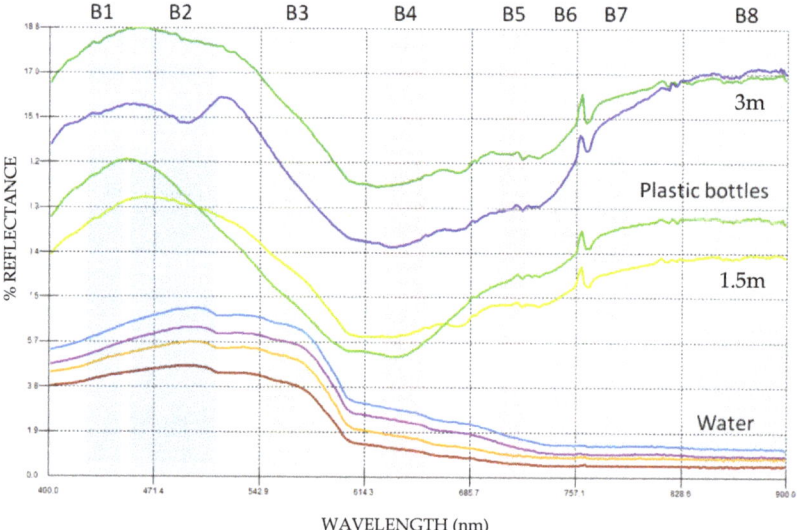

Figure 7. Average spectral signatures plotted against the different channels of the Sentinel-2 satellite. The red, orange, purple and cyan lines indicate the spectral signatures for water, while the light green, green, blue and dark green lines indicate the spectral signatures for the plastic bottles in the water at different heights (1.5 m and 3 m).

The above spectral signatures showed that the percentage reflectance at the infrared band was high for the plastic bottles, while that for the water was low. In the blue band, the percentage reflectance of both water and plastic bottles was high; hence, it was difficult to recognize the plastic bottles within the blue channel, as is evident in Figure 7. The plastic bottles had minimum reflectance value at B04 and maximum reflectance value at B08, which were also evident in the images taken with the infrared camera (Figure 6). Therefore, the images acquired from the Sentinel-2 satellite in the visible range did not detect the plastic litter target. In Figure 8, a yellow circle indicates where the plastic litter target should be visible.

Figure 8. Sentinel-2 Satellite RGB image (15 December 2018). (Bands B04, B03, B02).

Figure 9 presents the satellite images from the Sentinel-2 MSIL1C satellite overpass on 15 December 2018 as processed by the indices that were examined in order to identify the plastic litter target using the Sentinel-2 satellite images. In the results of the processed indices in Figure 9, all the above indices identified the plastic litter at the values featured in Figure 9, except for the AWEI. It was found that indices with B04 and B08 had better results since the spatial resolution was 10 m.

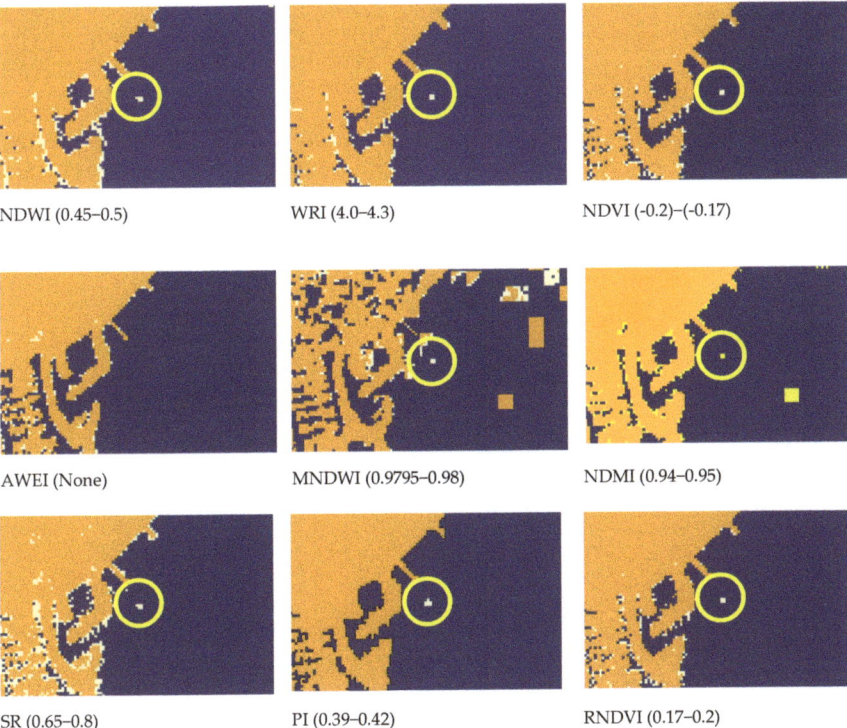

Figure 9. Sentinel-2 Satellite image (15 December 2018) processed with the indices described in the text with the corresponding values for which the plastic was detected. Land is represented in orange, water is represented in blue and plastic is represented in yellow. The yellow square within the yellow circle is the plastic litter target during the satellite overpass which corresponds to the Plastic Index Values (PIV), which is in the parentheses for each index.

The authors used the proposed statistical method of objective sensitivity analysis [96] to validate which index has the best capability of detecting plastic litter in water using the selected indices. The sensitivity analysis (SAV) is calculated using the discriminative value (DV) multiplied by the number of pixels detected (NPD) in the area of interest around the target, which is divided by the maximum value (Dmax) minus the minimum value (Dmin) detected for each index. The equation to identify the Sensitivity Analysis Values is expressed as:

$$SAV = (DV * NPD)/(Dmax - Dmin) \quad (10)$$

The sensitivity analysis values are shown in Table 2.

Table 2. Values used for sensitivity analysis.

Indices	Index MIN Value (Dmin)	Index MAX Value (Dmax)	Number of Plastic Pixels Detected (NPD)	Plastic Index Value (PIV)	Water Index Value (WIV)	Discriminative Value (DV)	Sensitivity Analysis Value (SAV)
NDWI	0.4612	0.6049	5	0.45–0.50	0.50–0.60	0.1	3.4795
WRI	4.0135	5.749	4	4.0–4.3	4.3–6.25	0.1	0.2305
NDVI	−0.3428	−0.1712	4	−0.2–−0.17	−0.37–−0.2	0.1	2.3310
AWEI	1241	2160	0	-	1241–2160	0	0.0000
MNDWI	0.9722	0.9797	4	0.97–0.98	0.96–0.97	0.0002	0.1067
MDMI	0.893	0.9459	4	0.86–0.94	0.94–0.95	0.1	7.5614
SR	0.4894	0.7076	5	0.65–0.8	0.45–0.65	0.1	2.2915
PI	0.3285	0.4143	7	0.39–0.42	0.31–0.37	0.2	16.3170
RNDVI	0.1712	0.3428	4	0.17–0.2	0.2–0.35	0.1	2.3310

Figure 10 (left) features the values where the plastics were detected, as determined in the sensitivity analysis using all of the nine indices. The sensitivity analysis performed on the indices indicated that the most optimal index to identify the plastic litter target was the PI (Equation (8)).

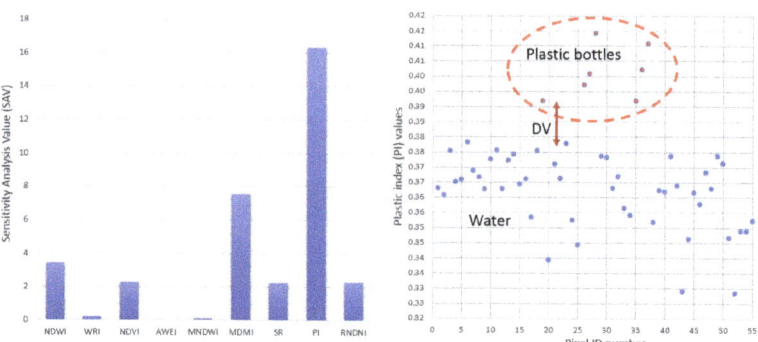

Figure 10. Left: results of the sensitivity analysis in bar graph for each of the nine indices. **Right**: Plastic Index (PI) scatterplot showing the discriminating value (DV) between water and plastic bottles. The plastic bottles values are circled in red.

The sensitivity analysis performed on the indices indicated that the most optimal index to identify the plastic litter target was the PI (Equation (8)). The scatterplot (Figure 10, right) shows the plastic index values (PIV) from the Plastic Index (PI) equation showing the water values clustered in the bottom and the plastic litter values at the top, providing a clear separation between them (DV). Figure 11 depicts the plastic litter target within the dotted yellow circle that was identified using the PI equation.

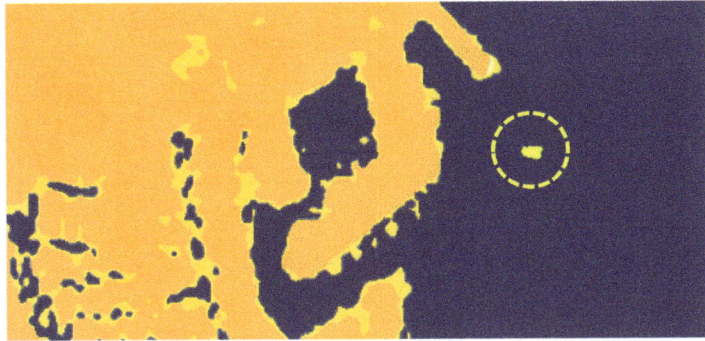

Figure 11. Plastic Index (PI) was used to identify the target, which is circled in yellow.

The PI equation was applied to the entire region of the Limassol coast, where the results identified the plastic litter target, as shown within the yellow dashed circle in Figure 12, as well as the fishing collars made of plastic and used in the floating fish farms off the coast of Limassol, which are encircled in orange. The yellow pixels encircled in blue in the upper right quadrant of the image give false positive values, as the Plastic Index identified them as plastic litter, although they were boats with plastic surfaces.

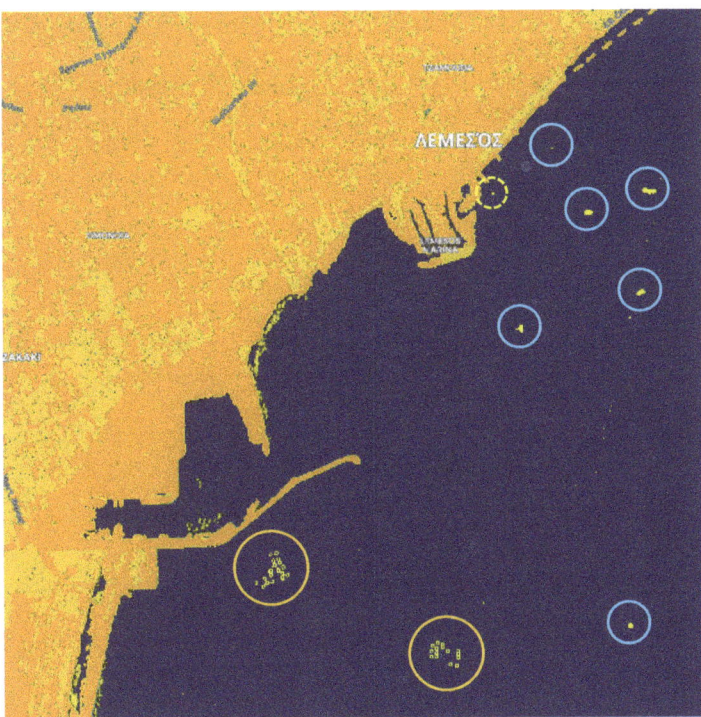

Figure 12. The PI applied to the coast of Limassol. The plastic litter target is in the dashed yellow circle, the floating fish farms are circled in orange and the false positives are circled in blue.

4. Discussion

The significance of this study is to identify whether plastic litter targets under 10 m can be detected with Sentinel-2 satellite images. This study examined the use of the Plastic Index (PI) and the Reversed Normalized Difference Vegetation Index (RNDVI) in order to identify plastic litter in water. In this study, a 3 m × 10 m plastic litter target was used to test if plastics could be detected. The study clearly shows that the 10 m spatial resolution bands (B04 and B08) can identify a plastic litter target below the Sentinel-2 pixel size. Our study used a smaller target, thereby differing from other research which used Sentinel-2 images to identify larger targets of 10 m [29,80]. The Plastic Index developed in this study can discriminate and identify plastic targets of 3 m × 10 m. The sensitivity analysis found that the Reversed Normalized Difference Vegetation Index produced low values and therefore was unable to identify plastic litter targets in water. The results of the PI were applied for the coast of Limassol. There is evidence that the PI may be used to discriminate smaller plastic targets. As seen in Figure 12, the fishing collars used in the floating fish farms off the coast of Limassol were identified by the PI. These collars are made of plastic and have a pipe diameter ranging from 50–110 cm; however, the collars themselves may have a diameter of 80 m. Future work may include smaller sizes of plastic targets as well as different types of plastic at different depths. This will provide the opportunity to validate and calibrate the PI accordingly.

The use of multispectral cameras mounted on a UAV provided the ability to check the reflectance of the plastic litter at different wavelengths, which was useful to visually identify the plastic reflectance response at these wavelengths. The acquisition of in situ spectral signatures of the plastic litter target within the water provided the ability to identify and compare with the spectral bands of Sentinel-2. The plastic litter detection was only examined between 400–900 nm. In this range, the reflectance of plastic bottles is low in the red band, but increases in the infrared bands, where water absorbs all the solar radiation and has almost no reflectance in the near infrared band, as indicated by Hafeez et al. [68]. The size of the plastic litter target was purposely used to emulate a plastic litter cluster and to find out if such a cluster can be identified from the Sentinel-2 satellite, as was conducted in other studies [29,62,80]. Plastic bottles have higher reflectance values in Bands 6, 7 and 8 of Sentinel-2 satellite images.

Based on the research featured in the Introduction, plastic can form clusters that may reach up to kilometers in size [35,36]. As a result, the intention of the study was not to identify individual plastic items but to focus on small clusters which had not been studied in previous research. Several indices were used to identify plastic litter targets in the study area. The evaluation of the images found that the Plastic Index using bands B04 and B08 of the Sentinel-2 MSIL1C sensor, developed and put forward by the authors, was the most effective in identifying the plastic litter as well as the pipe rings of floating fish farms that are located in the vicinity of the study area. One of the limitations of using smaller targets to detect plastic litter may be the satellite pixel size [66]. For this reason, future research can replicate this study by using high resolution satellite sensors (e.g., Planet Imagery, WorldView, GeoEye) to identify smaller plastic litter targets.

5. Conclusions

The results of this study indicate that plastic bottles in the sea can best be identified using the PI index at B04 and B08 using Sentinel-2 satellite images and the results assessed with spectral signatures and aerial images acquired from a UAV. The study found that Sentinel-2 satellite images were effective in identifying plastic clusters in the sea. As well, the study found that plastics in the sea can be identified by the high reflectance of solar radiation at NIR wavelengths. Two new indices, the Plastics Index (PI) and the Reversed Normalized Difference Vegetation Index (RNDVI) were formulated for processing the satellite images in the effort to identify plastic litter in the water. The methodology can only be used during cloud-free conditions using Sentinel-2 MSIL1C, without performing atmospheric correction due to the water spectral absorption. The findings of this study are significant, since, as indicated by various researchers, the south-east Mediterranean faces a significant problem with plastic debris.

This study developed an index that can identify plastics in the sea, which will assist the process of monitoring plastics in the Mediterranean Sea. The newly developed Plastic Index may be applied to other types of plastic debris, such as plastic bags, marine debris from aquaculture and other material, which were used in other research studies. It is vital to note that future research will focus on using high resolution images to identify plastics.

The study lends itself to further research in several areas. In addition, along the lines of this study, further investigation can be conducted with various configurations of different materials, such as plastic bags, at different depths in the sea. Future research can also investigate the use of Sentinel-1 Synthetic Aperture Radar (SAR) images for the identification of plastic litter in the seas. Further research can examine different sized plastic litter to identify the sensitivity of Sentinel-2 images in detecting subpixel plastic clusters. One of the significant outcomes of the study is that smaller plastic targets under 10 m can be identified using the Sentinel-2 satellite images.

Author Contributions: Conceptualization, C.P. and K.T.; methodology, K.T. and C.P.; formal analysis, K.T.; writing—original draft preparation, K.T.; writing—review and editing, K.T. and S.M.; visualization, K.T.; supervision, C.P., K.T. and D.H.; project administration, C.P., K.T., S.M. and D.H. All authors have read and agreed to the published version of the manuscript.

Funding: This research received no external funding.

Acknowledgments: The methodology presented in this paper was developed within the activities of the 'ERATOSTHENES: Excellence Research Centre for Earth Surveillance and Space-Based Monitoring of the Environment'—'EXCELSIOR' project, that has received funding from the European Union's Horizon 2020 Research and Innovation program, under Grant Agreement No 857510 and from the Government of the Republic of Cyprus, through the Directorate General for the European Programmes, Coordination and Development (www.excelsior2020.eu). Through this project, the ERATOSTHENES Research Centre of the Department of Civil Engineering and Geomatics at the Cyprus University of Technology is to be upgraded to ERATOSTHENES Centre of Excellence (ECoE). The authors acknowledge the contributions of the students and teachers of St. Peter and Paul Lyceum in Limassol, the Cyprus Diver's Association, the Limassol and Germasogeia Municipalities. We would also like to thank their reviewers for their comments.

Conflicts of Interest: The authors declare no conflict of interest.

References

1. UNEP—NOAA. Honolulu Strategy—A Global Framework for Prevention and Management of Marine Debris. 2011. Available online: https://repository.library.noaa.gov/view/noaa/10789 (accessed on 21 March 2019).
2. Kershaw, P.J. *Marine Plastic Debris and Microplastics—Global Lessons and Research to Inspire Action and Guide Policy Change*; United Nations Environment Programme: Nairobi, Kenya, 2016; ISBN 978-92-807-3580-6.
3. Thompson, R.C.; Swan, S.H.; Moore, C.J.; vom Saal, F.S. Our plastic age. *Philos. Trans. R. Soc. B Biol. Sci.* **2009**, *364*, 1973–1976. [CrossRef]
4. Algalita Marine Research Foundation. Annual Report. 2008. Available online: https://algalita.org/ (accessed on 21 March 2019).
5. Barnes, D.K.A.; Galgani, F.; Thompson, R.C.; Barlaz, M. Accumulation and fragmentation of plastic debris in global environments. *Philos. Trans. R. Soc. Lond. Ser. B Biol. Sci.* **2009**, *364*, 1985–1998. [CrossRef] [PubMed]
6. Gall, S.C.; Thompson, R.C. The impact of debris on marine life. *Mar. Pollut. Bull.* **2015**, *92*, 170–179. [CrossRef] [PubMed]
7. Thompson, R.C.; Olsen, Y.; Mitchell, R.P.; Davis, A.; Rowland, S.J.; John, A.W.G.; McGonigle, D.; Russell, E. Lost at sea: Where is all the plastic? *Science* **2004**, *304*, 838. [CrossRef] [PubMed]
8. Walker, T.R.; Reid, K.; Arnould, J.P.Y.; Croxall, J.P. Marine debris surveys at Bird Island, South Georgia 1990–1995. *Mar. Pollut. Bull.* **1997**, *34*, 61–65. [CrossRef]
9. Walker, T.R.; Grant, J.; Archambault, M.C. Accumulation of marine debris on an intertidal beach in an urban park (Halifax Harbour, Nova Scotia). *Water Qual. Res. J. Can.* **2006**, *41*, 256–262. [CrossRef]
10. Derraik, J.G.B. The pollution of the marine environment by plastic debris: A review. *Mar. Pollut. Bull.* **2002**, *44*, 842–852. [CrossRef]
11. Sánchez, P.; Mercedes, M.; Saez, R.; Juan, S.; Muntadas, A.; Demestre, A. Baseline study of the distribution of marine debris on soft-bottom habitats associated with trawling grounds in the northern Mediterranean. *Sci. Mar.* **2013**, *77*, 247–255. [CrossRef]

12. Thompson, R.C. Plastic debris in the marine environment: Consequences and solutions. In *Marine Nature Conservation in Europe*; Krause, J.C., Nordheim, H., Bräger, S., Eds.; Federal Agency for Nature Conservation: Stralsund, Germany, 2006; pp. 107–115.
13. Munari, C.; Corbau, C.; Simeoni, U.; Mistri, M. Marine litter on Mediterranean shores: Analysis of composition, spatial distribution and sources in north-western Adriatic beaches. *Waste Manag.* **2015**, *49*, 483–490. [CrossRef]
14. Laist, D.W. Overview of the Biological Effects of Lost and Discarded Plastic Debris in the Marine Environment. *Mar. Pollut. Bull.* **1987**, *18*, 319–326. [CrossRef]
15. Ivar Do Sul, J.A.; Costa, M.F. The present and future of microplastic pollution in the marine environment. *Environ. Pollut.* **2014**, *185*, 352–364. [CrossRef] [PubMed]
16. Jambeck, J.R.; Geyer, R.; Wilcox, C.; Siegler, T.R.; Perryman, M.; Andrady, A.; Narayan, R.; Law, K.L. Plastic waste inputs from land into the ocean. *Science* **2015**, *80*, 768–771. [CrossRef] [PubMed]
17. Nelms, S.E.; Duncan, E.M.; Broderick, A.C.; Galloway, T.S.; Godfrey, M.H.; Hamann, M.; Lindeque, P.K.; Godley, B.J. Plastic and marine turtles: A review and call for research. *ICES J. Mar. Sci.* **2016**, *73*, 165–181. [CrossRef]
18. Pettipas, S.; Bernier, M.; Walker, T. A Canadian policy framework to mitigate plastic marine pollution. *Mar. Policy* **2016**, *68*, 117–122. [CrossRef]
19. UNEP. Marine Litter: An Analytical Overview. Intergovernmental Oceanographic Commission of The United Nations Educational, Scientific and Cultural Organisation. 2005. Available online: http://wedocs.unep.org/bitstream/handle/20.500.11822/8348/-Marine%20Litter%2c%20an%20analytical%20overview-20053634.pdf?sequence=3&isAllowed=y (accessed on 1 March 2019).
20. Law, K.L.; Moret-Ferguson, S.; Maximenko, N.A.; Proskurowski, G.; Peacock, E.E.; Hafner, J.; Reddy, C.M. Plastic accumulation in the North Atlantic Subtropical Gyre. *Science* **2010**, *329*, 1185–1188. [CrossRef]
21. Gallo, F.; Fossi, C.; Weber, R.; Santillo, D.; Sousa, J.; Ingram, I.; Nadal, A.; Romano, D. Marine litter plastics and microplastics and their toxic chemicals components: The need for urgent preventive measures. *Environ. Sci. Eur.* **2018**, *30*, 13. [CrossRef]
22. McKinsey Center for Business and Environment. *Stemming the Tide: Land-Based Strategies for a Plastic-Free Ocean*; McKinsey & Company and Ocean Conservancy: New York, NY, USA, 2015; Available online: https://oceanconservancy.org/wp-content/uploads/2017/04/full-report-stemming-the.pdf (accessed on 7 April 2019).
23. Eriksen, M.; Lebreton, L.C.M.; Carson, H.S.; Thiel, M.; Moore, C.J.; Borerro, J.C.; Galgani, F.; Ryan, P.G.; Reisser, J. Plastic Pollution in the World's Oceans: More than 5 Trillion Plastic Pieces Weighing over 250,000 Tons Afloat at Sea. *PLoS ONE* **2014**, *9*, e111913. [CrossRef]
24. Ellen Macarthur Foundation. The new Plastics Economy: Rethinking the Future of Plastics. Ellen Macarthur Foundation, Cowes. 2016. Available online: https://www.ellenmacarthurfoundation.org/publications/the-new-plastics-economy-rethinking-the-future-of-plastics (accessed on 5 April 2019).
25. Carlson, D.F.; Suaria, G.; Aliani, S.; Fredj, E.; Fortibuoni, T.; Griffa, A.; Russo, A.; Melli, V. Combining litter observations with a regional ocean model to identify sources and sinks of floating debris in a semi-enclosed basin: The Adriatic Sea. *Front. Mar. Sci.* **2017**, *4*, 78. [CrossRef]
26. Corbin, C.J.; Singh, J.G. Marine debris contamination of beaches in St. Lucia and Dominica. *Mar. Pollut. Bull.* **1993**, *26*, 325–328. [CrossRef]
27. Kubota, M. A mechanism for the accumulation of floating marine debris north of Hawaii. *J. Phys. Oceanogr.* **1994**, *5*, 1059–1064. [CrossRef]
28. Kubota, M.; Takayama, K.; Namimoto, D. Pleading for the use of biodegradable polymers in favor of marine environments and to avoid an asbestos-like problem for the future. *Appl. Microbiol. Biotechnol.* **2005**, *67*, 469–476. [CrossRef] [PubMed]
29. Biermann, L.; Clewley, D.; Martinez-Vicente, V.; Topouzelis, K. Finding plastic patches in coastal waters using optical satellite data. *Sci. Rep.* **2020**, *10*, 53–64. [CrossRef] [PubMed]
30. Napper, I.E.; Thompson, R.C. Marine plastic pollution: Other than microplastic. In *Waste*; Elsevier: Amsterdam, The Netherlands, 2019; pp. 425–442. [CrossRef]
31. D'Asaro, E.A.; Shcherbina, A.Y.; Klymak, J.M.; Molemaker, J.; Novelli, G.; Guigand, C.M.; Haza, A.C.; Haus, B.K.; Ryan, E.H.; Jacobs, G.A.; et al. Ocean convergence and the dispersion of flotsam. *Proc. Natl. Acad. Sci. USA* **2018**, *115*, 1162–1167. [CrossRef]

32. Möhlenkamp, P.; Purser, A.; Thomsen, L. Plastic microbeads from cosmetic products: An experimental study of their hydrodynamic behaviour, vertical transport and resuspension in phytoplankton and sediment aggregates. *Elem. Sci. Anth.* **2018**, *6*, 61. [CrossRef]
33. Brooks, M.T.; Coles, V.J.; Coles, W.C. Inertia influences pelagic sargassum advection and distribution. *Geophys. Res. Lett.* **2019**, *46*, 2610–2618. [CrossRef]
34. Thiel, M.; Hinojosa, I.A.; Joschko, T.; Gutow, L. Spatio-temporal distribution of floating objects in the german bight (North Sea). *J. Sea Res.* **2011**, *65*, 368–379. [CrossRef]
35. Lebreton, L.; Slat, B.; Ferrari, F.; Sainte-Rose, B.; Aitken, J.; Marthouse, R.; Hajbane, S.; Cunsolo, S.; Schwarz, A.; Levivier, A.; et al. Evidence that the Great Pacific Garbage Patch is rapidly accumulating plastic. *Sci. Rep.* **2018**, *8*, 4666. [CrossRef]
36. Chu, S.; Wang, J.; Leong, G.; Woodward, L.A.; Letcher, R.J.; Li, Q.X. Perfluoroalkyl sulfonates and carboxylic acids in liver, muscle and adipose tissues of black-footed albatross (*Phoebastria nigripes*) from Midway Island, North Pacific Ocean. *Chemosphere* **2015**, *138*, 60–66. [CrossRef]
37. Pasquini, G.; Ronchi, F.; Strafella, P.; Scarcella, G.; Fortibuoni, T. Seabed litter composition, distribution and sources in the Northern and Central Adriatic Sea (Mediterranean). *Waste Manag.* **2016**, *58*, 41–51. [CrossRef]
38. Andrady, A.L. Microplastics in the Marine Environment. *Mar. Pollut. Bull.* **2011**, *62*, 1596–1605. [CrossRef]
39. Corcoran, P.L.; Biesinger, M.C.; Grifi, M. Plastics and beaches: A degrading relationship. *Mar. Pollut. Bull.* **2009**, *58*, 80–84. [CrossRef] [PubMed]
40. Gregory, M.R.; Andrady, A.L. Plastics in the marine environment. In *Plastics and the Environment*; John Wiley & Sons, Inc.: Hoboken, NJ, USA, 2003; pp. 379–402.
41. Shah, A.A.; Hasan, F.; Hameed, A.; Ahmed, S. Biological degradation of plastics: A comprehensive review. *Biotechnol. Adv.* **2008**, *26*, 246–265. [CrossRef] [PubMed]
42. Gregory, M.R. Environmental implications of plastic debris in marine settings entanglement, ingestion, smothering, hangers-on, hitch-hiking and alien invasions. *Philos. Trans. R. Soc. B* **2009**, *364*, 2013–2025. [CrossRef] [PubMed]
43. Rochman, C.M.; Browne, M.A.; Underwood, A.J.; van Franeker, J.A.; Thompson, R.C. The ecological impacts of marine debris: Unraveling the demonstrated evidence from what is perceived. *Ecology* **2015**, *97*, 302–312. [CrossRef] [PubMed]
44. Thompson, R.C.; Moore, C.J.; vom Saal, F.S.; Swan, S.H. Plastics, the environment and human health: Current consensus and future trends. *Philos. Trans. R. Soc. B* **2009**, *364*, 2153–2166. [CrossRef]
45. Kwon, B.G.; Koizumi, K.; Chung, S.-Y.; Kodera, Y.; Kim, J.-O.; Saido, K. Global styrene oligomers monitoring as new chemical contamination from polystyrene plastic marine pollution. *J. Hazard Mater.* **2015**, *300*, 359–367. [CrossRef]
46. Zettler, E.R.; Mincer, T.J.; Amaral-Zettler, L.A. Life in the "Plastisphere": Microbial communities on plastic marine debris. *Environ. Sci. Technol.* **2013**, *47*, 7137–7146. [CrossRef]
47. Wilcox, C.; van Sebille, E.; Hardesty, B.D. Threat of plastic pollution to seabirds is global, pervasive, and increasing. *Proc. Natl. Acad. Sci. USA* **2015**, *112*, 11899–11904. [CrossRef]
48. Schuyler, Q.A.; Wilcox, C.; Townsend, K.A.; Wedemeyer-Strombel, K.R.; Balazs, G.; van Sebille, E.; Hardesty, B.D. Risk analysis reveals global hotspots for marine debris ingestion by sea turtles. *Glob. Chang. Biol.* **2016**, *22*, 567–576. [CrossRef]
49. Sherman, P.; van Sebille, E. Modeling marine surface microplastic transport to assess optimal removal locations. *Environ. Res. Lett.* **2016**, *11*, 014006. [CrossRef]
50. UN Environment (2016) Regional Seas Programme. *Regional Seas Strategic Directions (2017–2020)*; Regional Seas Reports and Studies No.201; UN Environment Regional Seas Programme: Nairobi, Kenya, 2016; Available online: www.unep.org/regionalseas (accessed on 25 April 2019).
51. Notten, P. *Addressing Marine Plastics: A Systemic Approach*; United Nations Environment Programme: Nairobi, Kenya, 2019; Available online: http://wedocs.unep.org/bitstream/handle/20.500.11822/26746/marine_plastics.pdf?sequence=1&isAllowed=y (accessed on 18 March 2019).
52. GESAMP. *Guidelines for the Monitoring and Assessment of Plastic Litter in the Ocean*; Joint Group of Experts on the Scientific Aspects of Marine Environmental Protection: London, UK, 2019; Available online: http://www.gesamp.org/publications/guidelines-for-the-monitoring-and-assessment-of-plastic-litter-in-the-ocean (accessed on 18 March 2019).

53. GESAMP. *Sources, Fate and Effects of Microplastics in the Marine Environment—A Global Assessment (Part 1)*; Joint Group of Experts on the Scientific, Aspects of Marine Environmental Protection, 2015. Available online: http://www.gesamp.org/publications/reports-and-studies-no-90 (accessed on 1 April 2019).
54. UNEP/MAP. *Report Meeting of MED POL Focal Points Mediterranean Action Plan*; MED POL, WG. 334/8; UNEP/MAP: Kalamata, Greece, 29 July 2009.
55. Cozar, A.; Sanz-Martin, M.; Marti, E.; Gonzalez-Gordillo, J.I.; Ubeda, B.; Galvez, J.A.; Irigoien, X.; Duarte, C.M. Plastic accumulation in the Mediterranean Sea. *PLoS ONE* **2015**, *10*, e0121762. [CrossRef]
56. Fossi, M.C.; Romeo, T.; Baini, M.; Panti, C.; Marsili, L.; Campani, T.; Canese, S.; Galgani, F.; Druon, J.N.; Airoldi, S.; et al. Plastic debris occurrence, convergence areas and Fin Whales feeding ground in the Mediterranean Marine Protected Area Pelagos Sanctuary: A modelling approach. *Front. Mar. Sci.* **2017**, *4*, 167. [CrossRef]
57. van Sebille, E.; Wilcox, C.; Lebreton, L.C.M.; Maximenko, N.A.; Hardesty, B.D.; van Franeker, J.A.; Eriksen, M.; Siegel, D.; Galgani, F.; Law, K.L. A global inventory of small floating plastic debris. *Environ. Res. Lett.* **2015**, *10*, 124006. [CrossRef]
58. Alomar, C.; Estarellas, F.; Deudero, S. Microplastics in the Mediterranean Sea: Deposition in coastal shallow sediments, spatial variation and preferential grain size. *Mar. Environ. Res.* **2016**, *115*, 1–10. [CrossRef]
59. Suaria, G.; Avio, C.G.; Mineo, A.; Lattin, G.L.; Magaldi, M.G.; Belmonte, G.; Moore, C.J.; Regoli, F.; Aliani, S. The Mediterranean Plastic Soup: Synthetic polymers in Mediterranean surface waters. *Sci. Rep.* **2016**, *6*, 37551. [CrossRef] [PubMed]
60. Pasternak, G.; Zviely, D.; Assaf, A.; Spanier, E.; Ribic, C. Message in a bottle—The story of floating plastic in the eastern Mediterranean Sea. *Waste Manag.* **2018**, *77*, 67–77. [CrossRef]
61. Coll, M.; Piroddi, C.; Steenbeek, J.; Kaschner, K.; Ben Rais Lasram, F.; Aguzzi, J.; Ballesteros, E.; Bianchi, C.N.; Corbera, J.; Dailianis, T.; et al. The biodiversity of the Mediterranean Sea: Estimates, patterns, and threats. *PLoS ONE* **2010**, *5*, e11842. [CrossRef]
62. Hecht, A.; Pinardi, N.; Robinson, A.R.; Hecht, A.; Pinardi, N.; Robinson, A.R. Currents, water masses, eddies and jets in the Mediterranean Levantine Basin. *J. Phys. Oceanogr.* **1988**, *18*, 1320–1353. [CrossRef]
63. Mansui, J.; Molcard, A.; Ourmières, Y. Modelling the transport and accumulation of floating marine debris in the Mediterranean basin. *Mar. Pollut. Bull.* **2015**, *91*, 249–257. [CrossRef]
64. Zambianchi, E.; Trani, M.; Falco, P. Lagrangian transport of marine litter in the Mediterranean Sea. *Front. Environ. Sci.* **2017**, *5*, 5. [CrossRef]
65. Driedger, A.; Dürr, H.; Mitchell, K.; Van Cappellen, P. Plastic debris in the Laurentian Great Lakes: A review. *J. Great Lakes Res.* **2015**, *41*, 9–19. [CrossRef]
66. Hafeez, S.; Wong, M.; Abbas, S.; Nichol, J.; Kwok, C. Detection and Monitoring of Marine Pollution Using Remote Sensing Technologies. In *Monitoring of Marine Pollution*; Fouzia, H.B., Ed.; IntechOpen: London, UK, 2018. [CrossRef]
67. Aoyama, T. Extraction of marine debris in the Sea of Japan using high-spatial-resolution satellite images. In *Remote Sensing of the Oceans and Inland Waters: Techniques, Applications, and Challenges*; Frouin, R.J., Shenoi, S.C., Rao, K.H., Eds.; Proceedings of SPIE; SPIE: Bellingham, WA, USA, 2016; Volume 9878, p. 987817.
68. Garaba, S.; Aitken, J.; Slat, B.; Dierssem, H.; Lebreton, L.; Zielinski, O.; Reisser, J. Sensing ocean plastics with an airborne hyperspectral shortwave infrared imager. *Environ. Sci. Technol.* **2018**, *52*, 11699–11707. [CrossRef] [PubMed]
69. Goddijn-Murphy, L.; Peters, S.; Van Sebille, E.; James, N.A.; Gibb, S. Concept for a hyperspectral remote sensing algorithm for floating marine macro plastics. *Mar. Pollut. Bull.* **2018**, *126*, 255–262. [CrossRef] [PubMed]
70. Maximenko, N.; Arvesen, J.; Asner, G.; Carlton, J.; Castrence, M.; Centurioni, L.; Chao, Y.; Chapman, J.; Chirayath, V.; Corradi, P.; et al. Remote Sensing of Marine Debris to Study Dynamics, Balances and Trends. In Community White Paper Produced at the Workshop on Mission Concepts for Marine Debris Sensing. 2016. Available online: https://pdfs.semanticscholar.org/fab0/8714d6f6b12bbd7645582f03b386b019cd14.pdf (accessed on 18 March 2019).
71. Pichel, W.G.; Veenstra, T.S.; Churnside, J.H.; Arabini, E.; Friedman, K.S.; Foley, D.G.; Brainard, R.E.; Kiefer, D.; Ogle, S.; Clemente-Colón, P.; et al. GhostNet marine debris survey in the Gulf of Alaska—Satellite guidance and aircraft observations. *Mar. Pollut. Bull.* **2012**, *65*, 28–41. [CrossRef] [PubMed]

72. Davaasuren, N.; Marino, A.; Boardman, C.; Alparone, M.; Nunziata, F.; Ackermann, N.; Hajnsek, I. Detecting microplastics pollution in world oceans using SAR remote sensing. In Proceedings of the International Geoscience and Remote Sensing Symposium (IEEE), Valencia, Spain, 22–27 July 2018; IEEE: Piscataway, NJ, USA, 2018; pp. 938–941. [CrossRef]
73. Howe, K.L.; Dean, C.W.; John Kluge, J.; Soloviev, A.V.; Tartar, A.; Shivji, M.; Lehner, S.; Shen, H.; Perrie, W. Relative abundance of Bacillus spp., surfactant-associated bacterium present in a natural sea slick observed by satellite SAR imagery over the Gulf of Mexico. *Elem. Sci. Anth.* **2018**, *6*, 8. [CrossRef]
74. Nazeer, M.; Nichol, J.E. Combining landsat TM/ETM+ and HJ-1 A/B CCD sensors for monitoring coastal water quality in Hong Kong. *IEEE Geosci. Remote Sens. Lett.* **2015**, *12*, 1898–1902. [CrossRef]
75. Khorram, S.; Cheshire, H.; Geraci, A.L.; La Rosa, G. Water quality mapping of Augusta Bay, Italy from Landsat-TM data. *Int. J. Remote Sens.* **1991**, *12*, 803–808. [CrossRef]
76. Baban, S.M.J. Detecting water quality parameters in the Norfolk Broads, UK, using Landsat imagery. *Int. J. Remote Sens.* **1993**, *14*, 1247–1267. [CrossRef]
77. Pattiaratchi, C.; Lavery, P.; Wyllie, A.; Hick, P. Estimates of water quality in coastal waters using multi-date Landsat Thematic Mapper data. *Int. J. Remote Sens.* **1994**, *15*, 1571–1584. [CrossRef]
78. Lim, J.; Choi, M. Assessment of water quality based on Landsat 8 operational land imager associated with human activities in Korea. *Environ. Monit. Assess.* **2015**, *187*, 384. [CrossRef]
79. Liubartseva, S.; Coppini, G.; Lecci, R.; Creti, S. Regional approach to modeling the transport of floating plastic debris in the Adriatic Sea. *Mar. Pollut. Bull.* **2016**, *103*, 115–127. [CrossRef] [PubMed]
80. Topouzelis, K.; Papakonstantinou, A.; Graba, S. Detection of floating plastics from satellite and unmanned aerial systems (Plastic Litter Project 2018). *Int. J. Appl. Earth Obs. Geoinf.* **2019**, *79*, 175–183. [CrossRef]
81. Hörig, B.; Kühn, F.; Oschütz, F.; Lehmann, F. HyMap hyperspectral remote sensing to detect hydrocarbons. *Int. J. Remote Sens.* **2001**, *22*, 1413–1422. [CrossRef]
82. Martin, C.; Parkes, S.; Zhang, Q.; Zhang, X.; McCabe, M.F.; Duarte, C.M. Use of unmanned aerial vehicles for efficient beach litter monitoring. *Mar. Pollut. Bull.* **2018**, *131*, 662–673. [CrossRef] [PubMed]
83. Moy, K.; Neilson, B.; Chung, A.; Meadows, A.; Castrence, M.; Ambagis, S.; Davidson, K. Mapping coastal marine debris using aerial imagery and spatial analysis. *Mar. Pollut. Bull.* **2017**, *132*, 52–59. [CrossRef] [PubMed]
84. Martínez-Vicente, V.; Clark, J.R.; Corradi, P.; Aliani, S.; Arias, M.; Bochow, M.; Bonnery, G.; Cole, M.; Cózar, A.; Donnelly, R.; et al. Measuring Marine Plastic Debris from Space: Initial Assessment of Observation Requirements. *Remote Sens.* **2019**, *11*, 2443. [CrossRef]
85. Ocean Conservancy. *30th Anniversary of International Coastal Cleanup*; Annual Report; Ocean Conservancy: Washington, DC, USA, 2016; 24p, Available online: https://oceanconservancy.org/wp-content/uploads/2017/04/2016-Ocean-Conservancy-ICC-Report.pdf (accessed on 15 April 2019).
86. Pasternak, G.; Zviely, D.; Ribic, C.; Assaf, A.; Spanier, E. Sources, composition and spatial distribution of marine debris along the Mediterranean coast of Israel. *Mar. Pollut. Bull.* **2017**, *114*, 1036–1045. [CrossRef]
87. Galgani, F.; Hanke, G.; Maes, T. Global Distribution, Composition and Abundance of Marine Litter. In *Marine Anthropogenic Litter*; Bergmann, M., Gutow, L., Klages, M., Eds.; Springer: Berlin, Germany, 2015; pp. 29–56.
88. García-Rivera, S.; Lizaso, J.L.S.; Bellido, J.M.B. Composition, spatial distribution and sources of macro-marine litter on the Gulf of Alicante seafloor (Spanish Mediterranean). *Mar. Pollut. Bull.* **2017**, *121*, 249–259. [CrossRef]
89. Rokni, K.; Ahmad, A.; Selamat, A.; Hazini, S. Water Feature Extraction and Change Detection Using Multitemporal Landsat Imagery. *Remote Sens.* **2014**, *6*, 4173–4189. [CrossRef]
90. McFeeters, S. The Use of Normalized Difference Water Index (NDWI) in the Delineation of Open Water Features. *Int. J. Remote Sens.* **1996**, *17*, 1425–1432. [CrossRef]
91. Shen, L.; Li, C. Water body extraction from Landsat ETM+ imagery using adaboost algorithm. In Proceedings of the 18th International Conference on Geoinformatics, Beijing, China, 18–20 June 2010. [CrossRef]
92. Rouse, J.W.; Haas, R.H.; Schell, J.A.; Deering, D.W. Monitoring vegetation systems in the Great Plains with ERTS. In Proceedings of the Third Earth Resources Technology Satellite—1 Symposium; NASA SP-351, Washington, DC, USA, 31 December 1973; pp. 309–317.
93. Feyisa, G.L.; Meilby, H.; Fensholt, R.; Proud, S.R. Automated Water Extraction Index: A new technique for surface water mapping using Landsat imagery. *Remote Sens. Environ.* **2014**, *140*, 23–35. [CrossRef]

94. Xu, H. Modification of normalised difference water index (NDWI) to enhance open water features in remotely sensed imagery. *Int. J. Remote Sens.* **2006**, *27*, 3025–3033. [CrossRef]
95. Wilson, E.H.; Sader, S.A. Detection of forest harvest type using multiple dates of Landsat TM imagery. *Remote Sens. Environ.* **2002**, *80*, 385–396. [CrossRef]
96. Ustinov, E.A. *Sensitivity Analysis in Remote Sensing*; Springer International Publishing: Berlin/Heidelberg, Germany, 2015; ISBN 978-3-319-15840-2. [CrossRef]
97. Masoumi, H.; Safavi, S.M.; Khani, Z. Identification and classification of plastic resins using near infrared reflectance spectroscopy. *Int. J. Mech. Ind. Eng.* **2012**, *6*, 213–220.
98. Hopewell, J.; Dvorak, R.; Kosior, E. Plastics recycling: Challenges and opportunities. *Phil. Trans. R. Soc. B* **2009**, *364*, 2115–2126. [CrossRef]

© 2020 by the authors. Licensee MDPI, Basel, Switzerland. This article is an open access article distributed under the terms and conditions of the Creative Commons Attribution (CC BY) license (http://creativecommons.org/licenses/by/4.0/).

Article

A New Approach for Understanding Urban Microclimate by Integrating Complementary Predictors at Different Scales in Regression and Machine Learning Models

Lucille Alonso * and Florent Renard

UMR CNRS 5600 Environment, City and Society, Department of Geography and Spatial Planning, University Jean Moulin Lyon 3, 69007 Lyon, France; florent.renard@univ-lyon3.fr
* Correspondence: lucille.alonso@univ-lyon3.fr

Received: 1 June 2020; Accepted: 13 July 2020; Published: 29 July 2020

Abstract: Climate change is a major contemporary phenomenon with multiple consequences. In urban areas, it exacerbates the urban heat island phenomenon. It impacts the health of the inhabitants and the sensation of thermal discomfort felt in urban areas. Thus, it is necessary to estimate as well as possible the air temperature at any point of a territory, in particular in view of the ongoing rationalization of the network of fixed meteorological stations of Météo-France. Understanding the air temperature is increasingly in demand to input quantitative models related to a wide range of fields, such as hydrology, ecology, or climate change studies. This study thus proposes to model air temperature, measured during four mobile campaigns carried out during the summer months, between 2016 and 2019, in Lyon (France), in clear sky weather, using regression models based on 33 explanatory variables from traditionally used data, data from remote sensing by LiDAR (Light Detection and Ranging), or Landsat 8 satellite acquisition. Three types of statistical regression were experimented: partial least square regression, multiple linear regression, and a machine learning method, the random forest regression. For example, for the day of 30 August 2016, multiple linear regression explained 89% of the variance for the study days, with a root mean square error (RMSE) of only 0.23 °C. Variables such as surface temperature, Normalized Difference Vegetation Index (NDVI), and Modified Normalized Difference Water Index (MNDWI) have a strong impact on the estimation model. This study contributes to the emergence of urban cooling systems. The solutions available vary. For example, they may include increasing the proportion of vegetation on the ground, facades, or roofs, increasing the number of basins and water bodies to promote urban cooling, choosing water-retaining materials, humidifying the pavement, increasing the number of public fountains and foggers, or creating shade with stretched canvas.

Keywords: air temperature; surface temperature; LiDAR; multiple linear regression; Landsat 8; urban heat island

1. Introduction

Climate change is a major current phenomenon with multiple environmental, social and economic consequences [1]. In urban areas, it exacerbates the urban heat island (UHI) phenomenon [2,3] which is characterized by a difference in temperature between an urban area and the surrounding rural areas. In this case, the temperature in urban areas is higher than in rural areas, particularly at night [4,5]. The factors that contribute to heat intensification and UHI can be explicated mainly by the surface factors linked to the substitution of water surfaces, vegetation cover, and wetlands by artificial areas, causing low evaporation and evapotranspiration [6–12]. Buildings made of low-albedo materials with high thermal inertia capture, stock, and discharge the heat trapped with a thermal lag of several hours depending on the size and type of buildings and the climate [13]. This result is combined with the effect caused by structures made of low albedo supplies with high thermal inertia, which absorb and accumulate heat. The intensification of heat can also be caused by morphological parameters related to urban roughness and the sky-view factor (SVF) [14–16]. Indeed, the roughness can cause a diminution of the wind speed and the SVF can reduce the release of heat during the night [2]. Finally, anthropogenic parameters such as industrial heat emissions, heating, transport, or air conditioning can contribute as well to heat intensification [17–20], the "cities consume 78 percent of the world's energy and produce more than 60 percent of greenhouse gas emissions. However, they account for less than 2 percent of the Earth's surface" [21].

These two climatic manifestations have consequences on the health of the inhabitants [22] and on the sensation of thermal discomfort felt in urban areas [23,24]. Moreover, the increase in heat waves is clearly demonstrated, whether we look at the duration, intensity, or frequency [25]. The effects of heat waves are overlaid on the microclimatic characteristics of urban environments [26,27], as well as on the increasing urbanization process of the population. This increasing urbanization has a significant impact on urban microclimates and leads to warmer temperatures in cities [28–31]. One of the effects of the combination of these events is an increase in the premature number of heat stress related deaths [32]. In this context, local public actors are trying to prevent and reduce the human risks potentially generated by an increase in heat waves. Knowing and understanding the effect of the urban heat island is a key requirement for smart and sustainable city design [33]. According to the US Department of Energy, the United States spends $10 billion annually on energy to reduce the urban heat island effect [34]. In addition, mitigating urban overheating is an important financial issue since every 1 °C increase in temperature leads to a 2% to 4% increase in electricity demand [35]. In some regions, this increase would even vary between 0.45% and 4.6%, which would correspond to an additional electrical penalty of about 21W per degree of temperature increase per person [36]. This difference in energy consumption between urban and rural areas is mainly due to the fact that the cooling load of urban buildings is 13% higher than that of similar buildings in rural areas [37]. Thus, this relationship between electricity consumption and temperature has been clearly established [38]. In addition, a study in Chicago showed that adding 10% ground cover, or planting about three trees per plot of land, reduces energy costs by about $50–$90 (about 45–80 euros) per year per home [39].

In addition, air temperature is a main variable in explaining environmental conditions, especially urban conditions. It is also involved in many important ecological processes such as actual and potential evapotranspiration, net radiation, or the distribution of species [40]. Thus, knowledge of air temperature at any point in the territory is increasingly in demand to feed quantitative models related to a wide range of fields, such as hydrology, ecology [41], or climatology [42–44].

Consequently, the comprehension of air temperature models is essential for multiple applications in hydrology, land-use planning, or public health. Accurate knowledge of temperatures is a necessity both for the environment and for health policies, particularly in urban areas, which can contribute to improved urban planning in the context of UHI mitigation, and the creation of urban cooling islands (UCIs).

Thus, it is necessary to estimate the air temperature at any point in a territory as well as possible. This knowledge is directly dependent on the density of the measurement network, especially in view of the current rationalisation of the network of fixed meteorological stations of Météo-France [45]. In France, there are only a few agglomerations with their own network of fixed meteorological stations, such as Rennes and Dijon [46,47]. The air temperature evolving on a metric scale, at less than 100 meters [48,49], a very dense measurements network is needed. However, this is not the case in Lyon, which is the study area. Consequently, this study proposes to model air temperature using traditionally used data, data from remote sensing by LiDAR (Light Detection and Ranging), or Landsat 8 satellite acquisition and data produced by mobile measurement. These mobile measurements are very useful, as there is not yet a network of fixed weather stations sufficiently developed in Lyon, as in most large conurbations. In addition, the use of information obtained from airborne sensors or satellites to observe the earth's surface from the sky or from space is a methodology that effectively evaluates the spatial distribution of land surface variables at the local and regional scales [50] and can be used for temperature modelling.

Urban air temperature can be estimated using different interpolation techniques such as spline [51,52] and interpolation kriging. More recently, modelling by regression [51] or by neural networks and other machine learning techniques has emerged [53]. Multiple studies have addressed this issue, either by using classical spatial interpolations (deterministic [54] or stochastic [55]) or by multiple regressions [42,50,56–59]. Previous air temperature modelling studies in urban areas are mostly based on measurements from fixed stations [42,47,52,60–62]. Studies involving modelling based on mobile measurements are less common [49,63]. Moreover, there have been none in the study area, whether they involve modelling from fixed stations or mobile measurements. Thus, this study has a double focus: to provide a first modelling of the air temperature of the territory using mobile measurements.

Most of the studies based on mobile measurements have been carried out using automobiles, for example in Portland (USA) [63], in Nancy (France) [44,64], in Los Angeles (USA) [65], in Hong-Kong [66], in Brno (Czech Republic) [67], or in Sfax (Tunisia) [68–70]. However, there are many inherent limitations of motorized transport. An increase in temperature may be observed when the car stops or slows down due to red lights or traffic jams. The proximity of other cars combined with the immobility of the vehicle may explain this. On the contrary, when the speed of the car increases, cooler temperatures may be observed due to the cooling of the speed of travel. Thus, this measurement method limits the route to be monitored because of the many one-way streets or pedestrian areas. Consequently, in this study, the choice to do bicycle measurements was made. These bicycle measurements already exist but are not so frequent. For example, they have been used in some areas such as in Rotterdam (Netherlands) [71], in Shenzhen (China) [72], in Ohio [73], in Utrecht (Netherlands) [74], and on foot in Vancouver (Canada) [49].

Moreover, the explanatory variables used for modelling air temperature are, in many cases, those commonly used such as latitude, longitude, altitude, and slope [60,75,76], or even the land use land cover [77]. Only some studies integrate some remote sensing data such as Difference Vegetation Index (NDVI) or Normalized Difference Moisture Index (NDMI) [49,63,78]. This study therefore proposes to reproduce, as well as possible, the conditions encountered in the field as a function of the morphological diversity very present in the urban environment using the largest possible sample of explanatory variables.

To summarize, the implications of this new approach for the understanding of urban micro-climates are fourfold. Firstly, mobile measurements to acquire air temperature are used on the second French conurbation, which has never been thermally modelled, despite marked thermal discomfort. Then, this air temperature will be modelled with a very large sample of explanatory variables, including classic topo climatic variables (altitude, longitude, latitude, slope, exposure, and so on), variables derived from the characteristics of the urban morphology (sky view factor, variation in the height of buildings, etc.), or variables linked to the occupation and nature of the soil (vegetation, moisture, water, bare soil, etc.). One of the special features of this study for the acquisition of explanatory variables is the use of very diverse but complementary techniques, notably through the use of LiDAR or the analysis of data produced by the Landsat satellite. In a third step, a buffer analysis by simple linear regression is performed to test the best calculation unit for each variable that could get the highest coefficient of determination. Finally, the three modelling methods are used and compared. The first two are a stepwise multiple linear regression and a partial least square linear regression, which has the advantage of better integrating the collinear variables. The last one is the random forest method which is a relatively recent machine learning technique.

Thus, this study proposes first to delimit the study area, then to address the data acquisition methods and statistical procedures, and finally to analyse the results. This last part allows us to discuss the contribution of each predictor variable to the modelling of air temperature and measurement error. This research is aimed at improving urban planning in the context of climate change and mitigation at UHI.

2. Materials and Methods

2.1. Lyon: A Study Area Characterized by a Considerable Urban Morphological Diversity

The area of interest chosen for this study is the urban heart of the city of Lyon and part of the city of Villeurbanne, on the border with the 6th district of Lyon (Figure 1). This area has the advantage of grouping together a significant diversity of land use in an urban environment. It is mainly occupied by continuous urban fabric (50%) of which 12.3% is discontinuous dense urban fabric, as well as by industrial, commercial, military, or public units (19.5%). Water, roads (main and secondary), and vegetation cover also occupy a significant surface of the territory, with respectively, 7.3%, 14.3%, and 8.9% (Table 1).

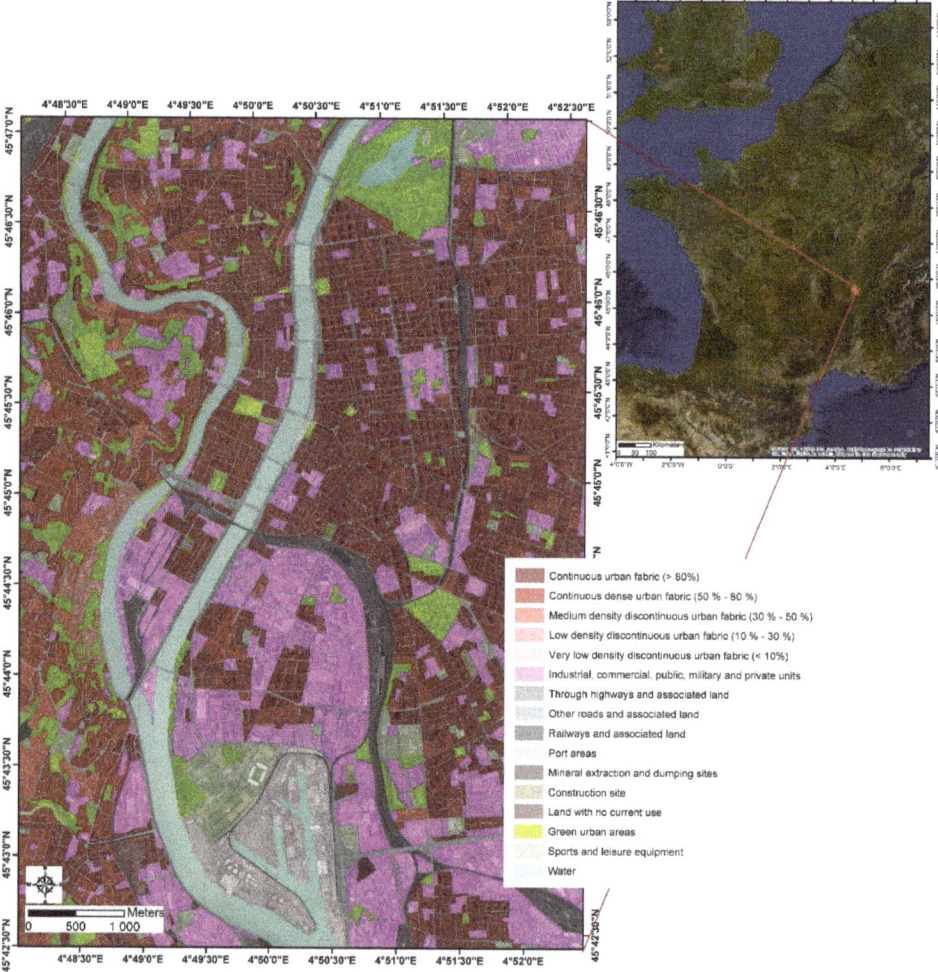

Figure 1. Location and land use of the study site (source: Urban Atlas 2012 and Data Grand Lyon).

Table 1. Land use/land cover distribution in the study area.

Land Use/Land Cover	Covered Surface Area (%)
Continuous urban fabric	50
Industrial, commercial, military, or public units	19.5
Roads (main and secondary)	14.3
Vegetation	8.9
Water	7.3

With just over 1.4 million inhabitants, this agglomeration of 59 municipalities is the second largest in France after Paris. The study area is composed of a very dense urban environment (Figure 1). Natural vegetation is therefore absent. There is, however, a very large park of 117 ha and urban green areas. The main park in Lyon (the Tête d'Or Park, to the north in Figure 1) is the largest urban park in France. It has vast expanses of lawn shaded by tall trees of various species, a lake, an island, and several botanical gardens, including an alpine garden and a flower garden.

This study area is located in the south-east of France (45°45′35″N, 4°50′32″E). According to the Köppen–Geiger classification [79,80], it has a continental temperate climate, fully humid with hot or warm summer, depending on the year (Figure 2). The hottest months are July and August, with average maximum temperatures of 27.7 °C and 27.2 °C, respectively. The wettest months are May and October with 90.8 mm and 98.6 mm, respectively. The sunniest months are June, July, and August with 254.3 h, 283 h and 252.7 h, respectively (Figure 2).

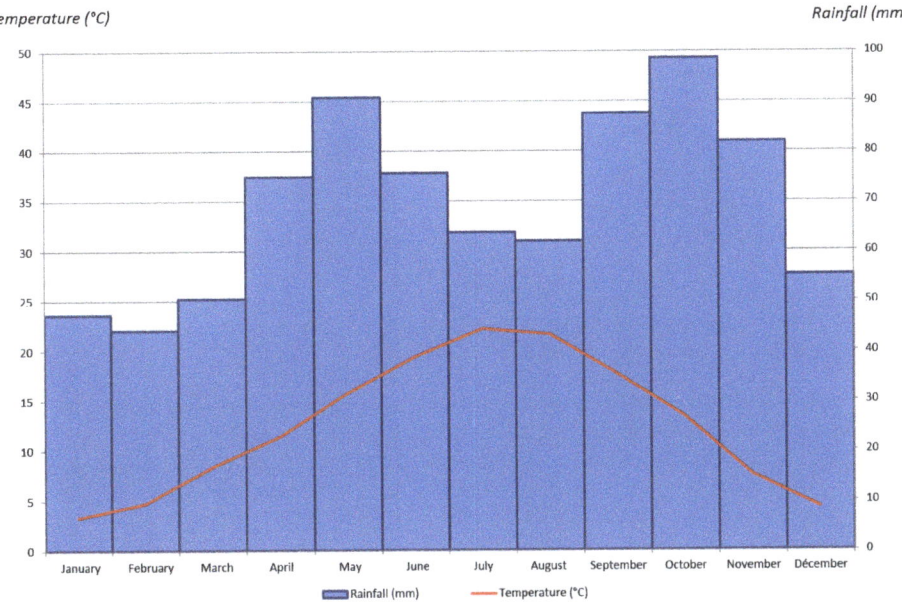

Figure 2. Ombrothermal diagram of Lyon (years 1981–2010; data: Météo-France).

2.2. Data Acquired by the Measuring Instruments and Selected Days

Air temperature is the variable to be estimated at any point in the territory from several indicators. The training sample of this variable is obtained from mobile measurement transects using high-precision measuring devices, according to manufacturer's data. The first equipment used is the EL-USB-1-RGC (EasyLog From Lascar Electronic). It measures the air temperature continuously, with an accuracy of +/− 1 °C (manufacturer's data) and a minimum recording interval of 1 second. The second equipment, the LOG 32 (from Dostmann electronic GmbH), records relative humidity and air temperature, with an accuracy of +/− 0.5 °C (manufacturer's data) and +/− 3% (40 to 60%) and a minimum recording interval of 2 seconds. The measurement campaigns were associated with a precision GPS (from Garmin, with a high sensitivity GPS/GLONASS receiver and Quad Helix antenna) to record the geographical position of the measurement.

The location of the points of all measurement campaigns was checked and corrected if necessary, using a geographic information system (GIS), for example, to ensure that it was not located on the roof of a building. Indeed, the dense urban environment can interfere with the geolocation of the position. Streets in dense urban centers may be boxed in, with little visible sky.

In addition, the Météo-France site of the Direction Centre-Est (DIRCE) of Lyon-Bron, (45°43′30″N, 4°56′12″E and 197 m altitude), was used as a study site for a quality control campaign of the air temperature and relative humidity measuring instruments. This station was chosen because it is Météo-France's professional weather station closest to our measurement campaigns. Hourly measurements synchronous to the measurements of the Météo-France station were carried out from 28 June 2018 at 09:00 to 24 September 2018 at 14:00 (Figure 3). The comparison between site observation and mobile measurements have been done at the same time, on a single second during the exact precise hour.

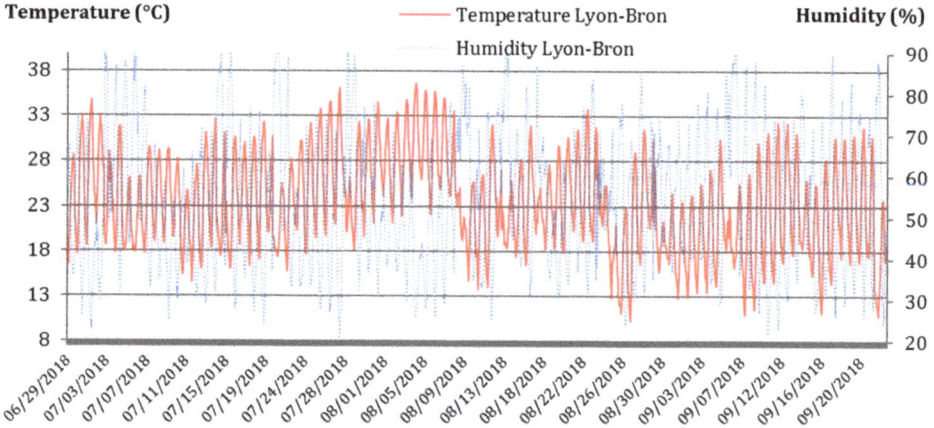

Figure 3. Hourly measurements of temperature (°C—red line) and humidity (%—blue dotted line) at the Météo-France Lyon-Bron station from 06/28/18 to 09/24/18.

The measuring devices used in this study proved to be highly accurate since, after this comparison at the Météo-France site over the summer 2018 period, the air temperature correlation of these two different acquisition sources shows a lowest correlation coefficient of 0.981 for the air temperature and 0.977 for the relative humidity measured by LOG 32 (Table 2).

Table 2. Synthesis of the correlation coefficients, root mean square error (RMSE) and MSE from the different measurement instruments used in relation to the Lyon-Bron station of Météo-France.

	LOG 32 n°1	LOG 32 n°2	EL-USB-1-RCG n°1	EL-USB-1-RCG n°2
Temperature (°C)	MSE: 0.892 RMSE: 0.944 R^2: 0.983	MSE: 0.797 RMSE: 0.893 R^2: 0.981	MSE: 0.516 RMSE: 0.718 R^2: 0.989	MSE: 0.566 RMSE: 0.752 R^2: 0.987
Humidity (%)	MSE: 12.305 RMSE: 3.507 R^2: 0.977	MSE: 11.970 RMSE: 3.459 R^2: 0.978		

The mobile measurements were taken on days when the Landsat 8 satellite was over the city, on clear sky days only, i.e., with less than 10% cloud cover. These campaigns were spread out between 2016 and 2018, exclusively over the summer period: 30 August 2016, 1 August 2017, 19 July 2018, and 22 July 2019. Numerous measurement campaigns were carried out from 2016 to 2019 but only those four days with similar weather conditions were used in this study. However, not all of them had similar weather conditions. Moreover, in only one summer of a year, the set of days was too poor regarding the different cumulative criteria, i.e., similar weather conditions and no clouds. Indeed, the weather conditions for each day were similar: the standard deviation of the temperature was only 0.9 °C and 4.3%, for humidity, 2.3 m.s-1 for wind speed, 132 degrees for wind direction, and 3.7 hPa for pressure. The average weather conditions for these indicators are 29.3 °C, 45.3%, 8.8 m.s-1, 260.8 degrees, and 1016.6 hPa, respectively (Table 3). Respectively, these measurement campaigns yielded 573, 300, 393, and 397 measurement points for air temperature and relative humidity (Figure 4).

Table 3. Meteorological parameters of the study days at the Lyon-Bron station at 12:00 noon (source: Météo-France).

	Temperature (°C)	Humidity (%)	Wind Speed (m/s)	Pressure (hPa)	Wind Direction (degrees)	Start	Finish
08/30/2016	27.7	46	9	1017.8	350	14:42	16:50
08/01/2017	29.4	52	10	1012.2	34	15:23	18:37
07/19/2018	29.8	42	5	1014.2	309	12:32	14:45
07/22/2019	30.1	41	11	1022	10	12:25	16:12
Mean	29.3	45.3	8.8	1016.6	260.8		
Standard deviation	0.9	4.3	2.3	3.7	132.0		
Minimum	27.7	41	5	1012.2	34		
Maximum	30.1	52	11	1022	350		

The routes travelled during the measurement campaigns vary slightly (Figure 4). In fact, besides the technical reasons such as works or new developments that caused us to deviate from the route, we wanted to maximize the morphological diversities crossed, making deviations to places of particular interest due to their urbanistic characteristics (docks, the historic urban center, industrial sectors, etc.).

Additionally, air temperature measurement campaigns sometimes last several tens of minutes. It was therefore necessary to make a correction based on a polynomial equation elaborated according to the evolution of the day's temperatures recorded at a time step of 10 minutes. This phase before the data processing is essential and allows to bring all these air temperature measurements back to the hottest hour of the day [74].

In addition, in order to have a very complete sample of temperature measurements, all the data from the four field trips were pooled. This allows obtaining global results. Indeed, even if the weather conditions are similar for the four days studied, some results may differ due to the different routes carried out, which cross different urban morphologies. For example, in Lyon, the type of buildings and the urban morphology are relatively different depending on whether one is in the east, with modern buildings from the end of the 19th and 20th centuries, or in the west of the main river, with very old buildings from the medieval or Renaissance period (Figures 1 and 4). This could explain why the results between the days of 2019 and 2018, for example, were not strictly identical, although general trends may emerge.

2.3. Morphological Descriptors Relevant to Air Temperature Estimation

Changes in land-use patterns related to the urban factory contribute to the spatial structuring of the urban landscape, which also influence energy transmission and balance [81,82]. These changes are considered a direct cause of the formation of the UHIs [83,84]. Thus, the relationship of changes in air temperature to land use and land cover is apparent.

Figure 4. Air temperature measurement points for 22 July 2019 (top left), 19 July 2018 (top right), 1 August 2017 (bottom left), and 30 August 2016 (bottom right—source: Data Grand Lyon).

In this study, thirty-eight explanatory variables contributed to the estimation of air temperature over the study area [85–89]. They belong to various categories such as climate data from remote sensing, topographic variables, vegetation indices, the presence of water, moisture, bare soil, buildings, radiation, urban morphology, and proximity to various land uses (Table 4 and Appendix A). The acquisition sources were multiple and came from the Landsat 8 satellite (https://earthexplorer.usgs.gov/), LiDAR (https://data.grandlyon.com/jeux-de-donnees/nuage-points-lidar-2018-metropole-lyon-format-laz/donnees) points and other cartographic products downloaded from the open data platform of the Greater Lyon.

Table 4. List of morphological descriptors used to estimate air temperature.

	Variables (Units)	Acquisition Source		Variables (Units)	Acquisition Source
Climate data from remote sensing	Surface temperature (°C)	Landsat 8	Building Index	NDBI Normalized Difference Built-Up Index	Landsat 8
	UTFVI Urban Thermal Field Variance Index	Landsat 8			
	Sunshine duration of the study day (h)	LiDAR data and modelling by ESRI ARCGIS		UI Urban Index	Landsat 8
	Radiation received for the study day (WH/m²)	LiDAR data and modelling by ESRI ARCGIS		IBI Index-based Built-Up Index	Landsat 8
Vegetation index	NDVI Normalized Difference Vegetation Index	Landsat 8		Building Density	LiDAR
	SAVI Soil Adjusted Vegetation Index	Landsat 8			
	EVI Enhanced Vegetation Index	Landsat 8			
	Tasseled Cap Transformation greenness (GVI)	Landsat 8	Topographic	Slope (°)	Data Grand Lyon
	Density of low vegetation	LiDAR			
	Density of medium vegetation	LiDAR		Exposure	LiDAR
	Density of high vegetation	LiDAR		Curvature	Data Grand Lyon
Water presence index	MNDWI Modified Normalized Difference Water Index	Landsat 8	Urban morphology	Sky View Factor	LiDAR
	NDWI Normalized Difference Water Index	Landsat 8		Standard Deviation (STD) of Building Height (building height variation)	Data Grand Lyon
Moisture index	Tasseled cap Transformation Wetness	Landsat 8	Land use	Distance to railway tracks	Data Grand Lyon
	NDMI Normalized Difference Moisture Index	Landsat 8		Distance to points of tourist interest	Data Grand Lyon
Bare soil index	NDBaI Normalized Difference Bareness Index	Landsat 8		Distance to subway entrances	Data Grand Lyon
	BI Bare Soil Index	Landsat 8		Distance to fountains	Data Grand Lyon
	EBBI Enhanced Built-Up and Bareness Index	Landsat 8		Water area	Data Grand Lyon
	Density of bare soil	LiDAR			
Radiation Index	Spectral radiance	Landsat 8			
	Emissivity	Landsat 8			
	Tasseled Cap Transformation Brightness	Landsat 8			

These morphological descriptors are acquired to a spatial precision that can go down to the centimeter scale. As a result, the information collected is dense and allows us to acquire a rigorous state of the urban environment for the purpose of modelling air temperature.

2.4. The Statistical Procedure Followed

2.4.1. An Explanatory Buffer Zone, Which Varies According to the Indicator

The aim of this study is to model air temperature using the linear regressions, multiple and partial least square regressions, and nonlinear regression by the random forest regression, from selected predictors. Initially, the scale with the best correlation between air temperature and explanatory variables was selected for each indicator based on a proximity buffer analysis (5 to 1000 m; Figure 5). Thus, the selected buffer zone varies for indicators of the presence of vegetation, water, humidity, bare soil and buildings, radiation indices, proximity to land use, urban morphology, and finally climate data (Table 5). Each of the measuring points was compared with the average of the indicator concerned, according to the size of the buffer considered.

For example, the process followed for the 5-meter buffer is as follows: 1°/creation of a 5-meter buffer around each point; 2°/calculation of the area (for vegetation, water surfaces, etc.), length (railways), average (spectral indices), or standard deviation (STD Building Height) of the indicator in this buffer; 3°/calculation of the Spearman correlation coefficient between the temperature measured at the point and the indicator; and 4°/repeat the operation for all the indicators and for all the buffers.

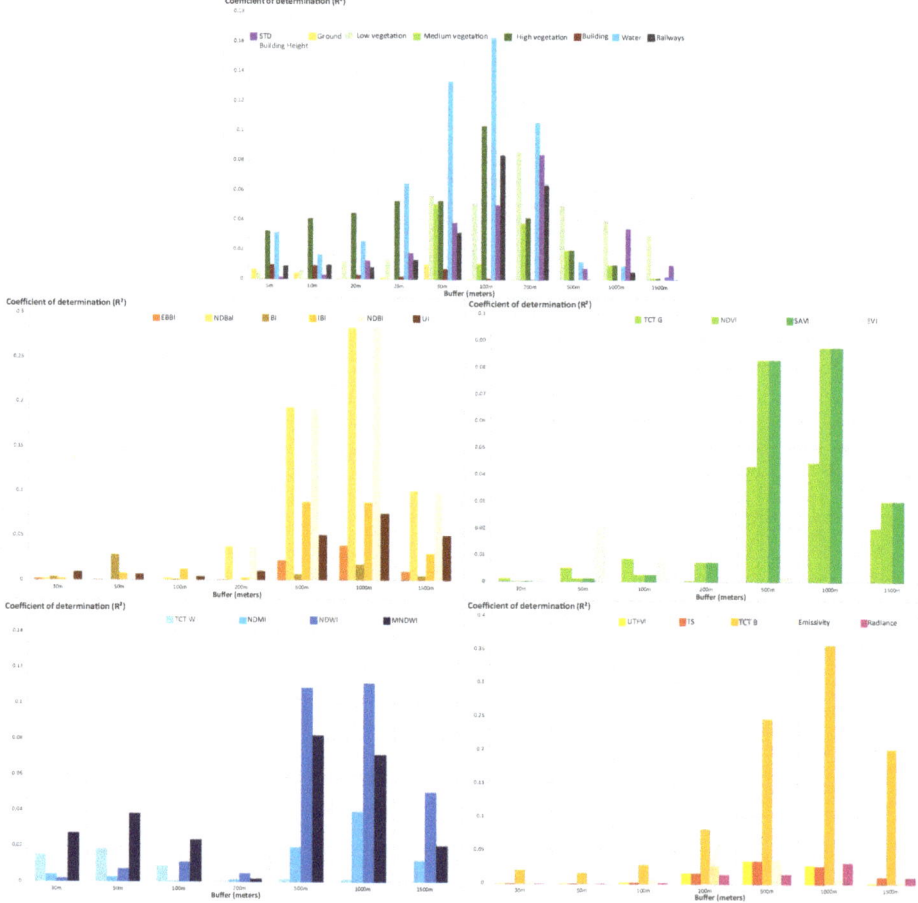

Figure 5. Example of variation in the correlation (coefficient of determination) between predictor and air temperature as a function of study scale.

Table 5. Buffer zones selected for each explanatory indicator.

	Variables (Unit)	Buffer Zone (m)		Variables (Unit)	Buffer Zone (m)
Climate data from remote sensing	Surface temperature (°C)	500	Bare soil index	NDBaI	1000
	UTFVI	500		BI	50
Vegetation index	NDVI	1000	Built index	EBBI	1000
	SAVI	1000		Density of bare soil	50
	EVI	50		NDBI	1000
	Tasseled Cap greenness	1000		UI	1000
	Density of low vegetation	200		IBI	500
	Density of medium vegetation	50		Density of built-up	5
	Density of high vegetation	100	Urban morphology	STD Building Height	100
Water index	MNDWI	500	Radiation Index	Spectral radiance	1000
	NDWI	500		Emissivity	500
Moisture index	Tasseled cap Wetness	50		Tasseled Cap Brightness	1000
	NDMI	1000	Land use	Density of water area	100

2.4.2. Three Complementary Regression Methods in Modelling Use

Three regression methods of air temperature modelling are compared in this study. These are two linear regressions: multiple [42,50,56] and partial least square [90], and one non-linear regression: random forest [91,92]. The aim is to select the best regression for this modelling. This evaluation is essentially carried out by comparing the coefficients of determination and the roots of the root mean square error (RMSE) obtained for the samples. The conditions of use for each of the regressions were also verified.

Multiple linear regression (MLR) is a data modelling method that requires several statistical steps before its application [93]. First, it is necessary to verify the normal distribution of the series in the dataset using the Shapiro–Wilk test (applies to samples of less than 5000 observations) [94]. This test has been invalidated, so the Spearman correlation matrix was used. It allows redundant variables not to be included in the regression model. One of the two indicators for which the pair has $|r|>0.7$ in the Spearman correlation matrix and a Variance Inflation Factor (VIF) > 5 was removed [95,96]. Finally, after removing the correlated variables, multiple linear regressions are carried out on about 20 variables, between 21 for the 30 August 2016, and 27 for the 1 August 2017 (Table 6). In addition, a holdout cross-validation was performed because of its ability to detect multiple regression overfitting (80% learning data and 20% validation data) [97].

In a complementary way to the multiple linear regression (MLR), partial least square regression is a method that is applied when a large number of explanatory variables are present and when these variables are likely to show strong collinearities among themselves [98]. Thus, this method allows us to model and predict air temperature values as a function of a linear combination of several quantitative (or qualitative) explanatory variables, overcoming the constraints of linear regression with respect to the distribution and number of variables included. Therefore, there is no need to remove the collinear variables. The model gives a value for each Variable Importance for the Projection (VIP). An explanatory variable is considered important when the VIP is greater than 0.8 [99]. A standardized coefficient is then generated for each of them [100].

Table 6. Non-collinear variables selected for multiple linear regressions per study day.

Variables	After Spearman Correlation Matrix and VIF			
	08/30/2016	08/01/2017	07/19/2018	07/22/2019
Surface temperature (°C)	X	X	X	X
UTFVI				
Sunshine duration of the study day				
Radiation received for the study day	X	X	X	
NDVI	X	X		X
SAVI				
EVI	X	X	X	X
Tasseled Cap greenness (GVI)				X
Density of low vegetation	X	X	X	X
Density of medium vegetation	X	X	X	X
Density of high vegetation	X	X	X	X
MNDWI	X	X	X	X
NDWI				X
Tasseled Cap Wetness	X	X	X	X
NDMI		X		
NDBaI			X	
BI	X	X	X	X
EBBI				
Density of bare soil	X	X	X	X
Spectral radiance				
Emissivity		X		X
Tasseled Cap Brightness	X	X		X
NDBI		X		
UI				
IBI		X		X
Building Density	X	X	X	X
Digital Elevation Model	X	X	X	X
Slope (°)	X	X	X	X
Longitude			X	
Exposure	X	X	X	
Curvature	X	X	X	X
Sky View Factor	X	X	X	X
STD Building Height	X			X
Distance to railway tracks	X	X	X	X
Distance to points of tourist interest		X	X	X
Distance to subway entrances		X	X	X
Distance to fountains		X	X	X
Water area	X	X	X	X
Final Number	21	27	22	26

The third type of regression tested is the random forest regression. This is a predictive model using binary decision trees [101]. From an observation sample, the bagging method will generate several possibilities before selecting only one. This machine learning technique [102] is based on Classification and Regression Trees (CART). These are constructed from different bootstrap samples, randomly selected with random discounting, in order to obtain, after aggregation, a robust and efficient set of air temperature predictors [103]. The importance of each variable is calculated by the mean increase in error of a tree in the forest, i.e., when the values of each variable are randomly swapped in the out-of-bag (OOB) samples. The variables used in the nonlinear regression, random forest, to model air temperature are derived from the selection of multiple linear regressions for each day. The random t forest classification and regression has the advantage of reducing white noise, and thus potentially improving the correlation coefficients and RMSE already obtained by multiple linear regression. In addition, the number of variables in the bagging and the number of trees used are user-defined parameters. When the number of trees increases, the general error converges to the same value. Overfitting is then not a problem due to the large numbers law. Despite this, the number of analysed trees must be limited in order not to excessively increase the computation time (1):

$$c \times T \times v \times (M \times N \times \log N) \tag{1}$$

where c is a constant, T is the number of trees in the set, M is the number of variables, and N is the number of samples in the training data set [104]. In this work, the classifiers were optimized with 80 decision trees and were trained with the same number of pixels in each category. The general error of the models converged around 80 decision trees (Figure 6). Therefore, a more complex model would have required more computation time without improving the classification.

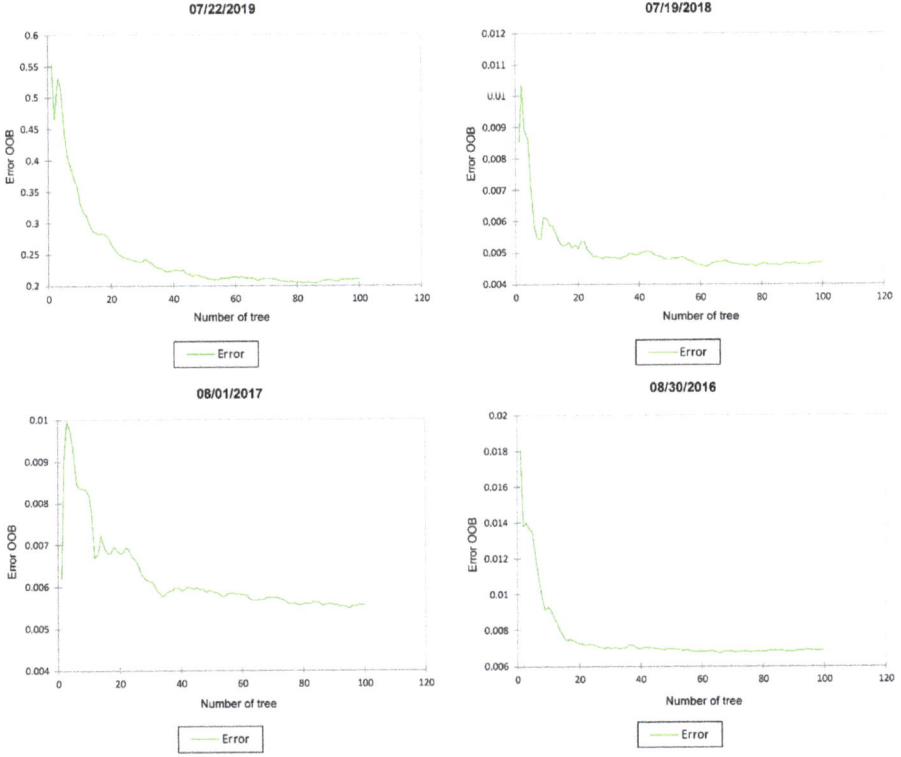

Figure 6. Convergence of the general error of the models for each study days.

In addition, Lasso regression was not applicable in this study. Lasso regression is only used when the number of predictors is greater than the number of observations [105,106]. Here, though, the number of observations was much higher than the number of predictors. In addition, many explanatory variables were included.

2.4.3. Quality Control on Modeling by Spatial Identification of Error Clusters

The spatial autocorrelation of the difference between the modelled air temperature and the air temperature measured by the mobile measurements was analyzed, on one hand, using the Anselin

Local Moran I spatial association indicator (LISA) [107], and, on the other hand, using the degree of clustering of high and low intensity values by the Getis Ord General G (Gi*) [108,109].

The LISA makes it possible to group together, for statistically significant results ($p < 0.05$), the similarity of a spatial unit with its neighbours. It allows identifying spatial aggregates of entities with high or low values as well as spatial outliers. A cartographic representation showing a cluster type for each statistically significant entity is thus obtained. With a geographic information system (GIS), a statistically significant group of high values (HH), a group of low values (LL), an outlier in which a high value is surrounded mainly by low values (HL), and an outlier in which a low value is surrounded mainly by high values (LH) is distinguished.

The local application of the general G statistic is the Getis Ord Gi* statistic. It is used to identify statistically significant ($p < 0.05$) spatial clusters of high and low intensity. Thus, for positive Z scores, the higher the Z score, the stronger the cluster of high intensity values (error overestimating air temperature). On the contrary, the lower the negative Z-score, the higher the group of low intensity values (error underestimating the air temperature).

In order to summarize the methodology, a general diagram of the study has been inserted (Figure 7).

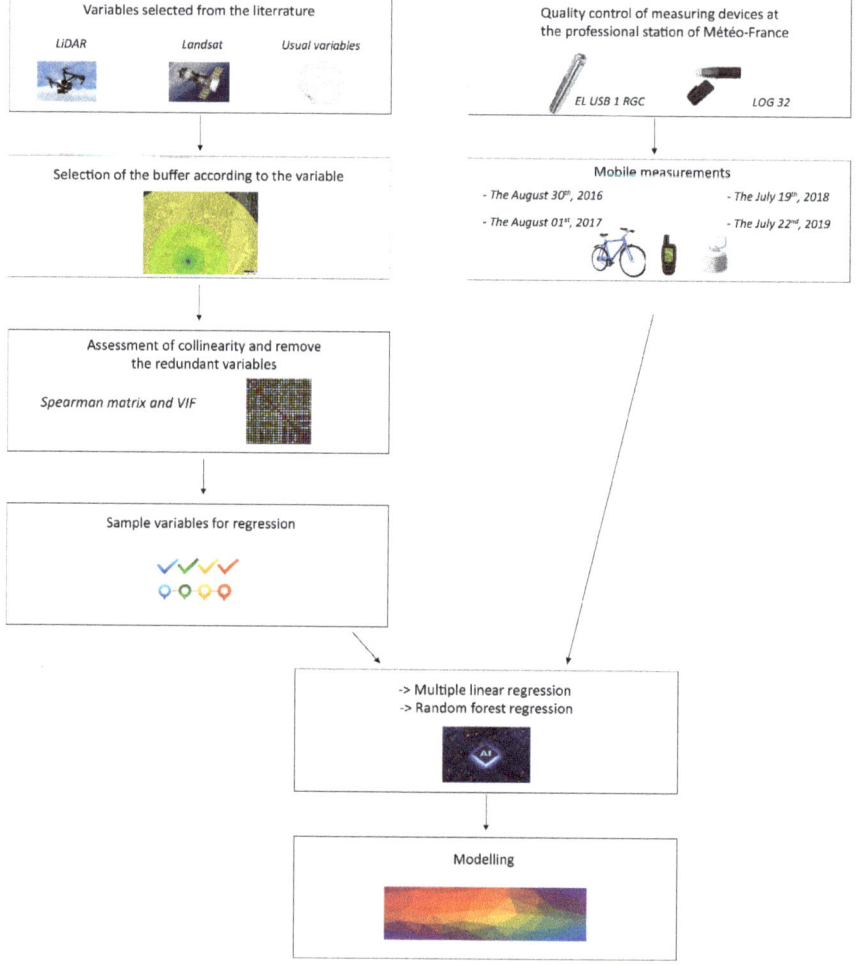

Figure 7. The methodological framework used for air temperature modeling.

3. Results

3.1. Multiple Linear Regression Modeling

After removing the collinear variables for each day, the predictors involved in air temperature modelling provide significant coefficients of determination. These ranged from 0.60 for 22 July 2019, to 0.89 for 30 August 2016, with RMSEs of only 0.96 °C and 0.23 °C, respectively. Moreover, each variable retained in the model was characterized by a normalized coefficient that corresponded to the weight of this explanatory variable. This weight varies according to the study days (Figure 8). The probability associated with Fisher's F (Pr>F) was always less than 0.05 and very often less than 0.0001.

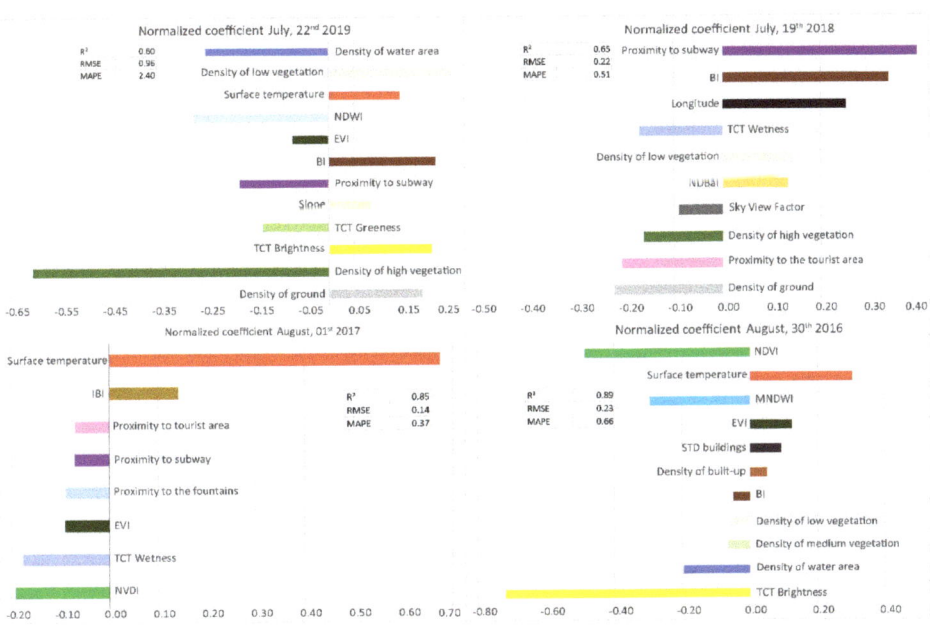

Figure 8. Weights of selected variables for each of the study days.

For example, for 19 July 2018, the variables contributing to a positive impact on the model were proximity to subways, Bare Soil Index (BI), longitude, Tasseled Cap Transformation (TCT) wetness, low vegetation density, and Normalized Difference Bareness Index (NDBaI). Variables negatively impacting the model are sky view factor, high vegetation density, proximity to tourist attractions, and soil density. From the equation obtained, it is therefore possible to model the air temperature continuously. The resolution can be adapted to the display and the purpose of the study. For example, a resolution of 10 meters was chosen for Figure 9 (Figure 9). It can be seen in Figure 9 that some areas are cooler or hotter than others on the map. This is directly related to the equation used in the modelling, including the explanatory variables included, as shown in Figure 8. Thus, for example, in Figure 9, cold spots at some locations are related to the density of tall vegetation or water surface. Hot spots, in contrast, would be related to low vegetation density, soil density, or BI. Thus, the greater the presence of these variables, the greater the chance of detecting a hot or cold spot. This confirms the results of previous studies showing in particular the cooling power of tall vegetation [43,110], water surfaces, or urban density [33,111,112].

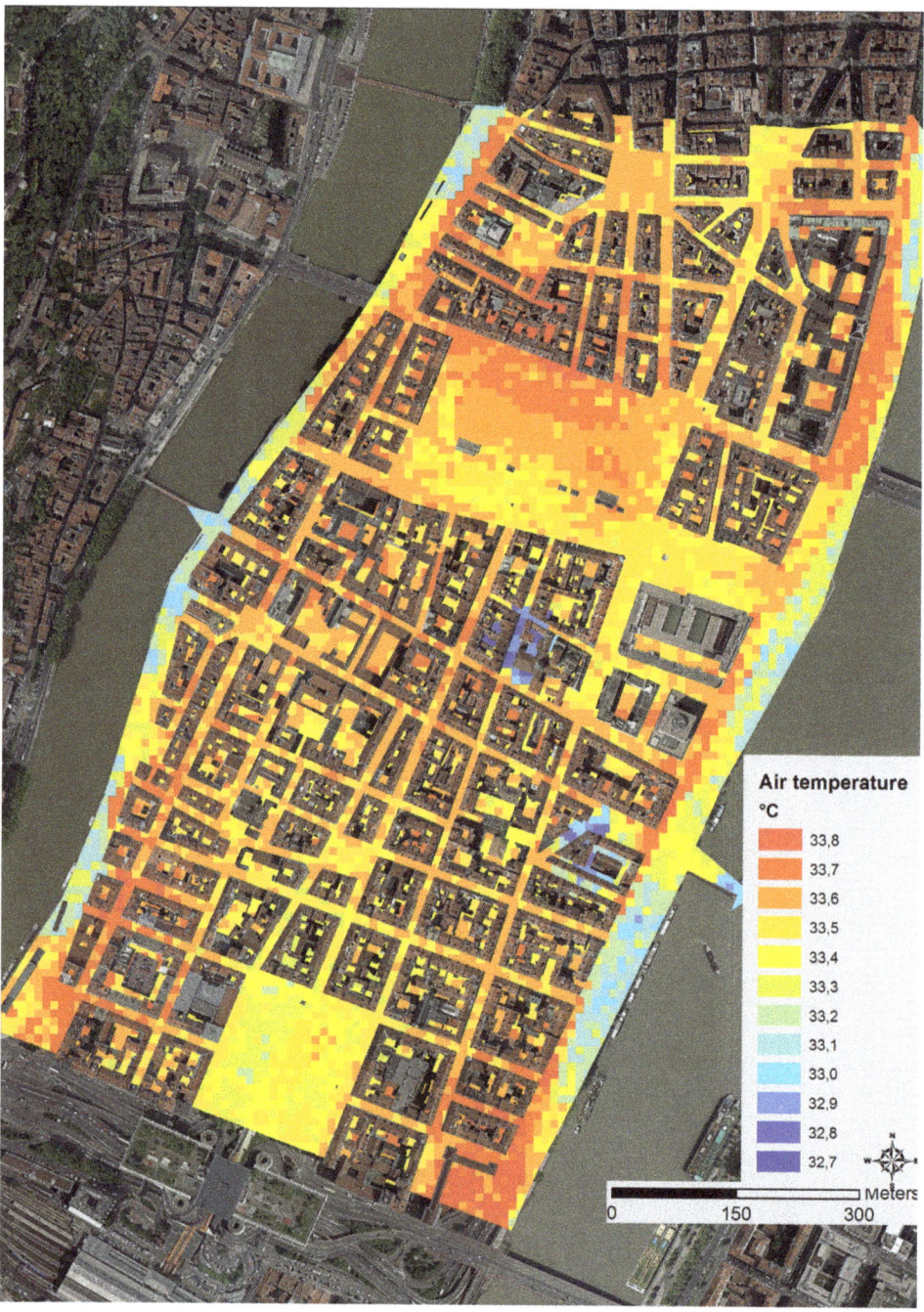

Figure 9. Modelling of air temperature in the dense urban center of Lyon on 19 July 2018 (source: Data Grand Lyon).

The results for the sample with all measurements for the four outputs (Figure 10) show an R^2 of 0.65 and an RMSE of 1.54. The results for the sample with all measurements for the four outputs (Figure 10) show an R^2 of 0.65 and an RMSE of 1.54. The RMSE is logically slightly higher than for the single day models due to the larger sample of measurements and greater morphological diversity, even though the weather conditions remain similar. These results confirm the general trends observed at day scales. In particular, the cooling effect of variables such as water density (normalized coefficient of −0.35; Figure 10), densely vegetated areas (−0.11 for NDVI and −0.09 for the density of high vegetation), road embankment (−0.08 for SVF), and humidity (−0.05 for Modified Normalized Difference Water Index (MNDWI)) can be found. The presence of proximity to tourist areas can be explained by the fact that these areas are mostly made up of green spaces or historic buildings in old Lyon.

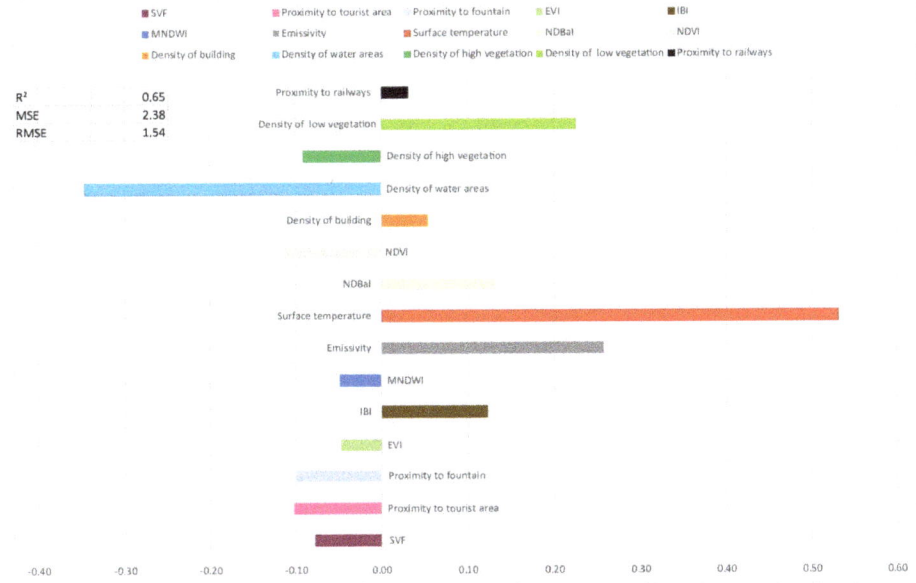

Figure 10. Weights of selected variables for the global sample on the four dates.

The variables contributing to urban warming were logically surface temperature and emissivity (normalized coefficients of 0.53 and 0.26, respectively; Figure 10) as well as indicators of the built environment and the absence of medium or high vegetation density (0.13 for the NDBaI, 0.12 for the Index-based Built-Up Index (IBI), and 0.23 for the density of low vegetation).

3.2. Partial Least Square Regression Modeling

Partial least square (PLS) regression modeling did not show much consistency in air temperature prediction since the mean coefficient of determination for all four study days is only 0.62, with a maximum of 0.79 for 30 August 2016, and a minimum of 0.53 for 22 July 2019 (Table 7). In addition, a large number of explanatory variables were retained, with a maximum of 26 for the day of 22 July 2019. Some variables influenced the model both positively and negatively as a function of the day. For example, the MNDWI had a significant positive impact for 30 August 2016, and a negative impact for 1 August 2017, 19 July 2018, and 22 July 2019. As a result, the air temperature modeling results were much less relevant than by multiple linear regression.

Overall PLS modelling based on the measurements of the four outputs provided results relatively similar to multiple linear regression, with an R^2 equal to 0.699 and an RMSE of 1.503. They also confirmed the dominant role of surface temperature. This variable had a VIP of 2.2. This is followed by the density of water areas (VIP = 1.81), the density of low vegetation (1.43), the NDVI (1.15), and the humidity indices (1.03 for the MNDWI and 1.01 for the TCT Wetness). These results are in agreement with those obtained through multiple linear regression (Section 3.1).

Table 7. Statistical parameters of the three explanatory variables used in air temperature modelling by partial least square linear regression.

Date	R^2	MSE	RMSE	Variables	Model Parameter in Absolute Value	Impact on the Model
08/30/2016	0.79	0.11	0.33	LST	0.0675	Negative
				NDVI	1.71	Positive
				MNDWI	4.53	Positive
08/01/2017	0.77	0.03	0.18	BI	0.58	Positive
				NDMI	0.51	Negative
				NDBI	0.51	Positive
07/19/2018	0.37	0.09	0.07	Emissivity	2.1128	Negative
				Longitude	1.3906	Positive
				NDBaI	1.2262	Positive
07/22/2019	0.,53	1.13	1.06	Emissivity	7.4782	Positive
				BI	3.0472	Positive
				NDBaI	2.5931	Positive
Mean	0.62	0.34	0.41			

3.3. Random Forest Regression Modeling

For the four study days, the coefficients of determination obtained were strong: 0.98 for the 30 August 2016, 0.96 for the 1 August 2017, 0.95 for the 19 July 2018, and 0.92 for the 22 July 2019 (Table 8). Thus, on average, a coefficient of determination of 0.95 was obtained, with a RMSE of only 0.17 °C and an out-of-bag (OOB) error of 0.05.

Table 8. Summary of Coefficients of Determination, Out-Of-Bag Error and Root Mean Square Error of Random Forest Classification, and Regression Modeling Errors.

Date	R^2	Out-Of-Bag	RMSE
08/30/2016	0.98	0.0071	0.08
08/01/2017	0.96	0.0045	0.07
07/19/2018	0.95	0.0071	0.08
07/22/2019	0.92	0.19	0.44
Mean	0.95	0.05	0.17

In addition, the measure of importance for each of the variables was measured by the mean increase in error of a tree in the forest when the observed values of that variable were randomly swapped in the out-of-bag samples (OOB; Figure 11). As a reminder, an increase in errors allowed us to know the importance of the variable in the modeling.

Global random forest modelling based on all days highlighted the dominant role of surface temperature, which had a mean error increase of 102.5 (Figure 12). This was followed by emissivity, density of low vegetation, IBI, and density of high vegetation with mean error increases of 26.9, 23.5, 16.2, and 16.2, respectively (Figure 12). These results are in agreement with those obtained using multiple linear regression and PLS regression.

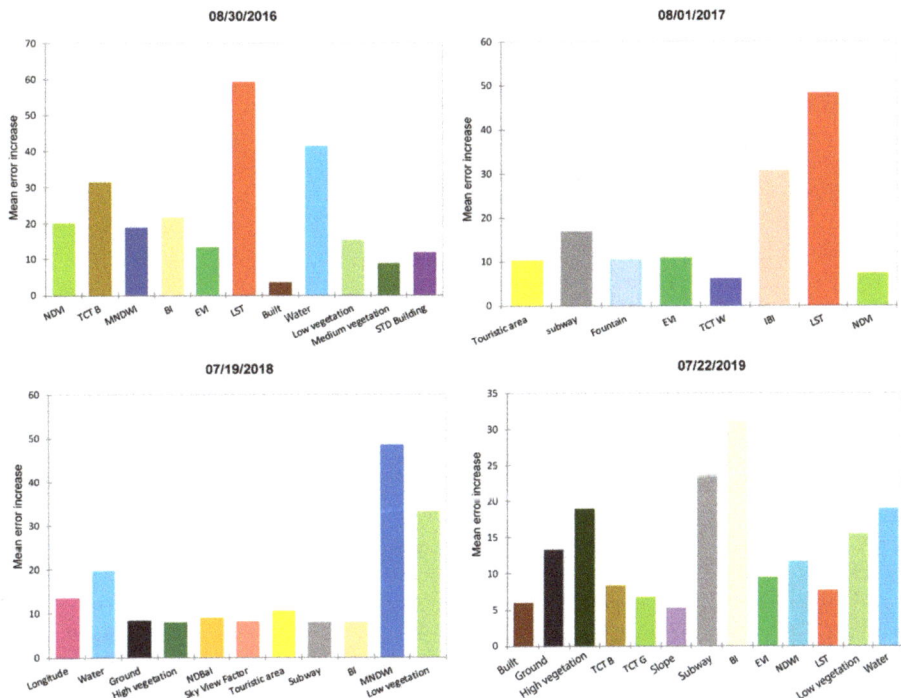

Figure 11. Evolution of the importance of the variables selected in random forest classification and regression modelling for the four study dates.

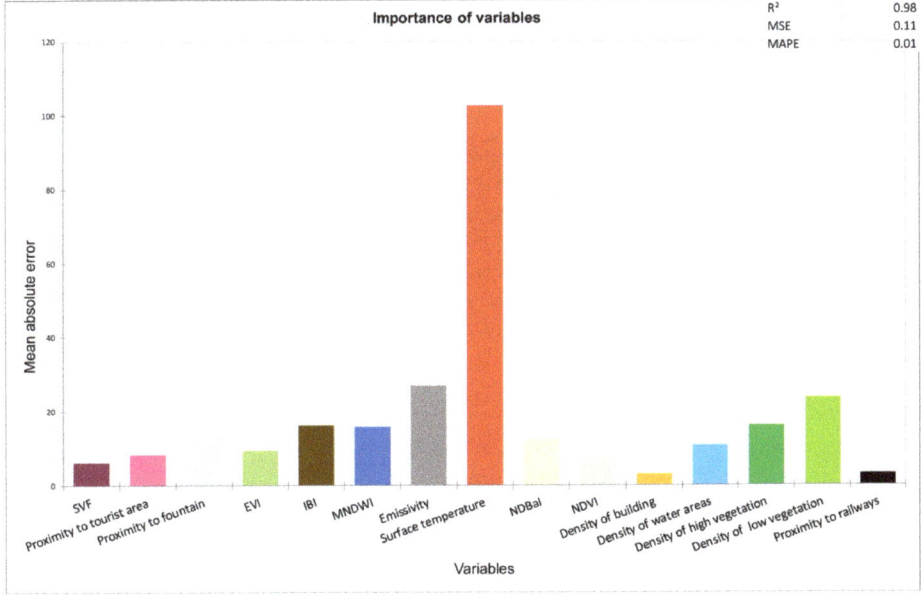

Figure 12. Evolution of the importance of the variables selected in random forest classification and regression modelling for the global modelling.

4. Discussion

4.1. Implication of Important Predictors in Urban Air Temperature Modeling

The results of the simple regressions (Section 2.4.1 and Figure 4), the multiple linear regression (Section 3.1), the Random forest regression (Section 3.3), and to a lesser extent the PLS regression (Section 3.2) make it possible to identify the parameters that positively and negatively influence urban air temperature. Naturally, surface temperature is frequently used in air temperature modelling (22 July 2019; 1 August 2017; and 30 August 2016; Figure 8) with very high weights (normalized coefficient of 0.67 for the day of 1 August 2017; Figure 8), mean error increase of 59.2 and 48.1 for the days of 30 August 2016, and 1 August 2017, respectively (Figure 11, for example). This is confirmed by the overall results of multiple linear regression and random forest modelling. Its normalized coefficient is 0.53 and its mean error increase is 102.5. However, this is not an urban morphological descriptor on which designers, urban planners, or politicians can directly act. The urban parameters highlighted by the models that can be influenced in planning operations to combat extreme temperatures in cities mainly concern green and blue solutions, and grey solutions [113–115].

Modelling results indicate that the factors that contribute to increasing temperatures in urban areas are related to building density. Indeed, regarding the density of buildings, for example, the BI had a normalized coefficient of 0.22 on 22 July 2019, and 0.34 on 19 July 2018. The TCT Brightness, which refers to bare, partially covered or waterproofed soils (such as rocky outcrops, concrete, gravel, asphalt, etc.) had a normalized coefficient of 0.21 for 22 July 2019, and a mean error increase of 31.5 for 30 August 2016, and 8.4 for 22 July 2019. The IBI had a coefficient of 0.14 and a mean error increase of 30.5 for 1 August 2017. To a lesser extent, we also found the presence of low vegetation (and thus the absence of high vegetation), which had normalized coefficients of 0.22 and 0.15 for the days of 22 July 2019 and 19 July 2018, respectively, and mean error increases of 15 and 35 for 30 August 2016, and 19 July 2018, respectively. This was confirmed by the overall results of multiple linear regression and random forest modelling. Its normalized coefficient was 0.23 and its mean error increase is 23.5 (Figures 10 and 12).

By contrast, the factors that favor the decrease in urban temperatures were related to the presence of vegetation, humidity, and surface water. Thus, high vegetation density had a high cooling power in the models with normalized coefficients of −0.6 for 22 July 2019 and −0.15 for 19 July 2018, and a mean error increase of 19 for 22 July 2019. These results are consistent with those of the global modelling using multiple linear regression and random forest. Its normalized coefficient is −0.09 and its mean error increase is 16.2 (Figures 10 and 12).

In addition, NDVI was found with normalized coefficients of −0.5, −0.18, and −0.11 and mean error increases of 20, 8, and 7.6 for the days of 2016, 2017, and globally, respectively, but also TCT greenness (−0.14 for 22 July 2019). Moisture also had a cooling effect through the TCT wetness (with normalized coefficients of −0.18 and −0.17 and mean error of 6 for 1 August 2017, and 19 July 2018, respectively) or proximity to fountains (with normalized coefficient of −0.09 and mean error of 10.4 for 1 August 2017). Finally, the density of water area has a negative impact on the model with normalized coefficients of −0.29, −0.20, and −0.35, and mean error increases of 18.9, 41.4, and 10.7, respectively, for 22 July 2019, 30 August 2016, and in a global way, as well as the NDWI (with a normalized coefficient of −0.27 and mean error of 11.7 for 22 July 2019) or the MNDWI (with respectively for 30 August 2016, and 19 July 2018, a mean error of 19 and 48.6).

To our knowledge, this is one of the studies aimed at modelling air temperature in a morphologically contrasting urban environment, with areas of unequally dense habitat, two rivers, a historic center and the largest urban park in France (Figure 1), which uses the most extensive sample of explanatory variables, with classic data, LiDAR data, and data from remote sensing. This study has shown the interest of the complementary use of the latter two types of data, in particular LiDAR for a precise view of vegetation densities (high, medium, or low) but also remote sensing for surface temperature and water, humidity, and vegetation indices to a lesser extent. While we expected very satisfactory results with random forest modelling, confirming the results of previous studies [63,97,116], we were surprised to note also the very high performance of the classical multiple linear regression and PLS regression, with very low RMSE and often below 0.5 °C.

These results confirm the roles played by vegetated spaces [43,110], building density, and water surfaces in previous studies [33,111,112] and confirm mitigation practices based on green and blue space solutions [113]. This study also highlighted the relatively low cooling power of low vegetation during the sunny afternoons of the measurement campaigns. Even if we had observed this during the measurement campaigns, this had yet to be confirmed by the models. This weakness can be explained by the density of vegetation, which is not high enough to promote sufficient evapotranspiration for cooling, but above all by a lack of shade compared to the tall vegetation.

Finally, the influence of buildings on air temperature is generally considered within a radius of 500 m [117–120]. However, in this study, it was found that the buffers with the best correlation between air temperature and building density, based on LiDAR data, were 5 and 10 m. Furthermore, the vegetation density obtained from LiDAR, which explains air temperature in an optimum way, was within a radius of 50 to 200 m, regardless of its height (low, medium, or high; Figure 5). This also corresponds to a smaller buffer size than that used in previous studies [46]. On the other hand, the buffer size for the bare-soil surfaces was between 50 m and 1000 m depending on the indicator (respectively density of bare-soil and BI, and NDBaI, and Enhanced Built-Up and Bareness Index (EBBI)). In this latter case, this size is similar to that of previous studies [46].

Consequently, this methodology based on mobile cycling measures, buffer analysis and regressions using complementary explanatory variables are fully applicable to other cities. However, we recommend testing the choice of scales for these variables, all the more so if it is not an old European city with morphological urban similarities to Lyon, although the optimal radii found in this study coincide with previous similar experiences in other cities.

In addition, it can be noted that similar spectral indices significantly correlate with the air temperature at the same scale because of the similar physical meaning represented by the indices. For example, vegetation indices such as TCT greenness, NDVI, and Soil Adjusted Vegetation Index (SAVI) are most relevant between 500 and 1000 m, as are water indices (NDWI or MNDWI), building indices (NDBI and Urban Index UI) and bare soil indices (NDBaI and TCT Brightness).

4.2. Spatialization of Error

The modelling error found is minimal for multiple linear regression and random forest modelling. For all study days, the median for multiple linear regression modelling was 0.02 °C and for random forest classification and regression modelling was 0.002 °C (σ of 0.44 and 0.17, respectively; Table 9). In contrast to the closeness of the median and mean values by these two modelling methods, the agreement is stronger for multiple linear regression than for random classification and regression forest (Figure 13).

Table 9. Multiple linear regression (MLR) and random forest regression (RDF) model error descriptive statistics.

	MLR	RDF
Biggest negative error (°C)	−2.23	−0.99
Biggest maximum error (°C)	2.50	1.29
First Quartile	−0.17	−0.05
Median	0.02	0.002
Third Quartile	0.17	0.05
Mean	0.01	0.003
Variance	0.19	0.03
Standard deviation	0.44	0.17

When looking at the location of errors in air temperature modelling between the multiple linear regression method and the random forest, similarities between the two are observed. The models overestimate the air temperatures towards the water areas on Confluence (south of the peninsula) and near the Perrache train station. They underestimate the air temperature in the streets in the embankment, near green areas, and south of the left bank of the Rhône (Figure 14).

Remote Sens. **2020**, *12*, 2434

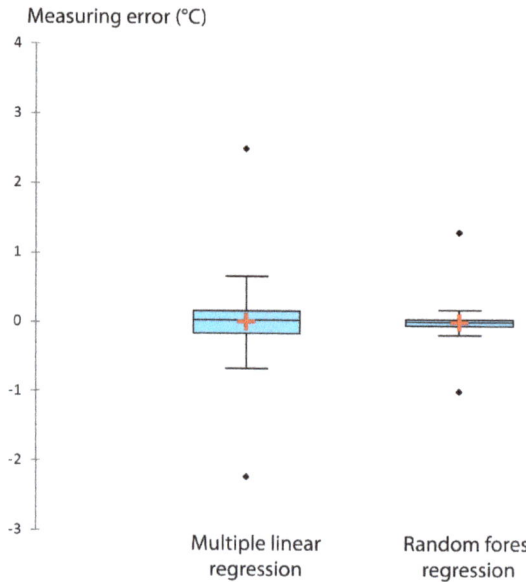

Figure 13. Box plots representation of the modeling error from multiple linear regression and random forest classification and regression.

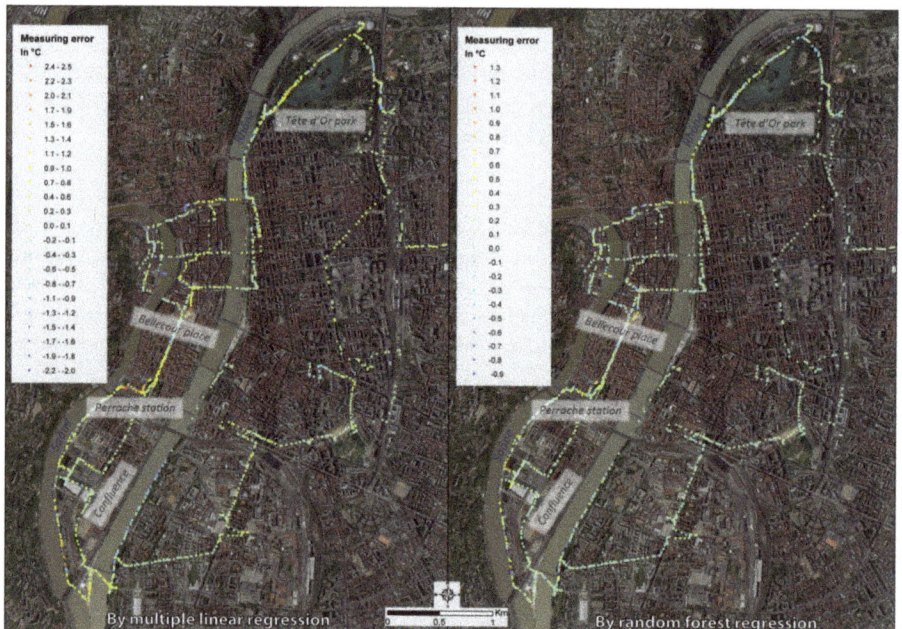

Figure 14. Location of the modeled measurement error of the air temperature by multiple linear regression (left) and by random forest (right) for all the study days (source: Data Grand Lyon).

If we analyze the location of these errors day by day, we notice that for the 30 August 2016, the multiple linear regression model overestimates the air temperatures near the water areas, on the bridges and south of the left bank of the Rhône. Conversely, it underestimates this physical magnitude on open spaces such as Bellecour Place. For 1 August 2017, 19 July 2018, and 22 July 2019, the model overestimates the air temperature also near the waterways and on the bridges but also on open spaces. In contrast, the model suggests that the streets in the embankments are cooler than in the mobile in situ measurements (Figure 15). The same can be seen in the random forest modelling (Figure 16). In addition, an overestimation of air temperature near the green spaces of the Tête d'Or Park was observed for the days of 30 August 2016, 19 July 2018, and 22 July 2019.

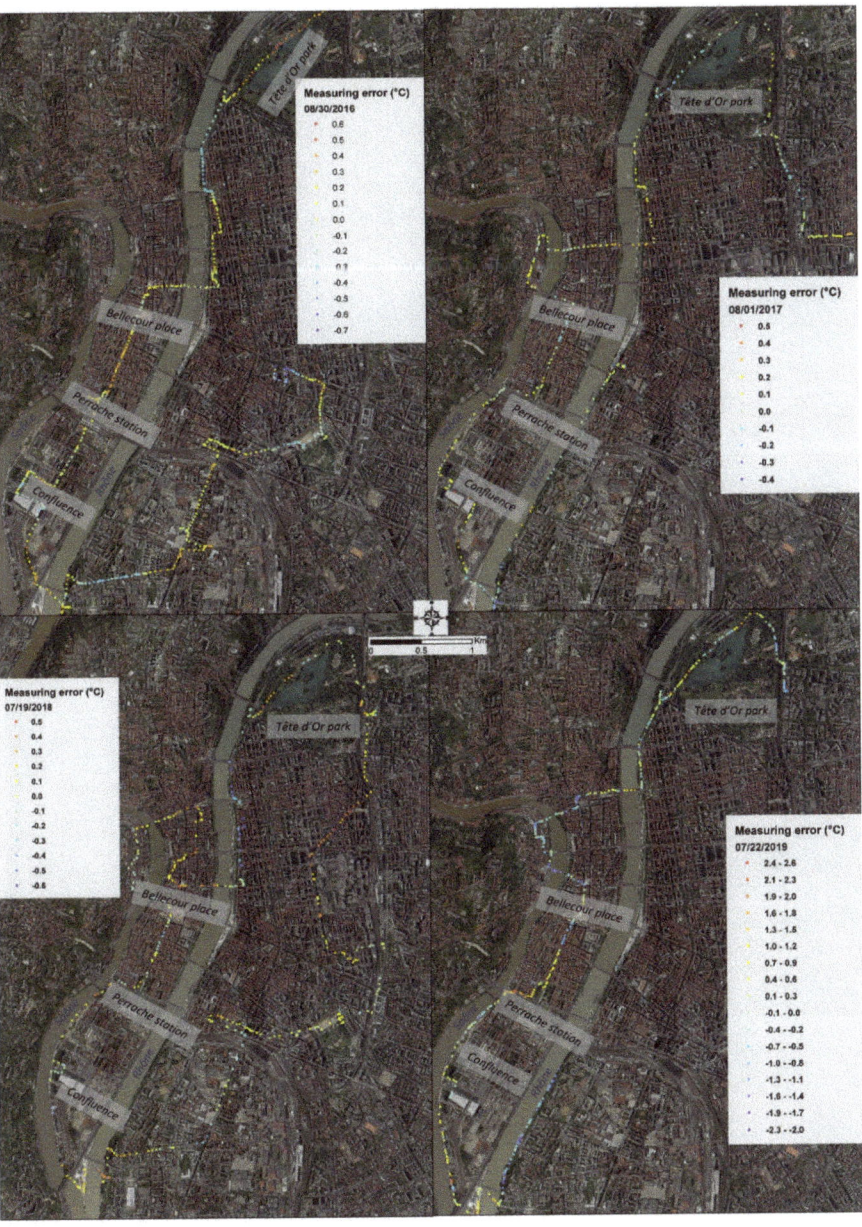

Figure 15. Localization of the measurement error of the air temperature modeling by multiple linear regression (source: Data Grand Lyon).

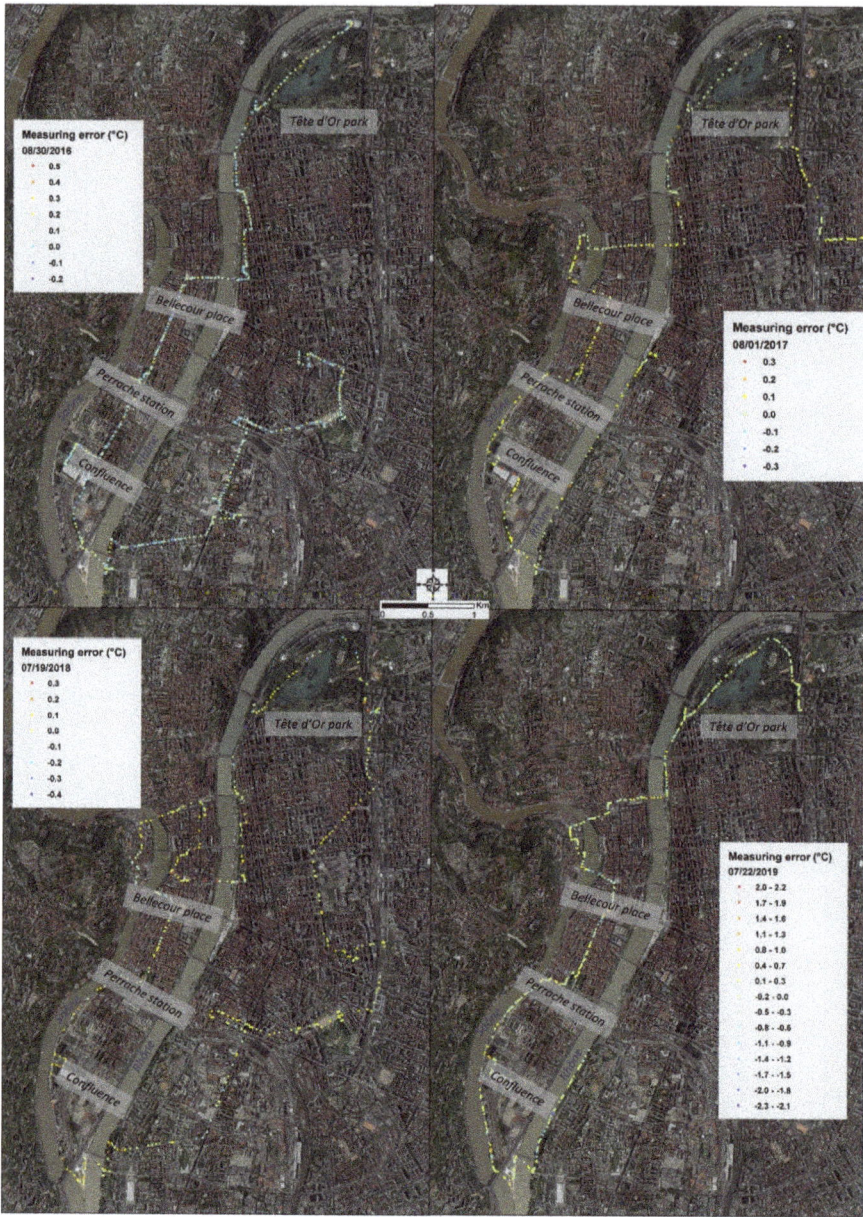

Figure 16. Localization of the measurement error of the air temperature modeling by random forest regression (source: Data Grand Lyon).

4.3. Grouping of Similar Errors

Spatial groupings of statistically similar values of the differences between modelled and measured air temperatures are evaluated using LISA (Figure 17) and Gi* (Figure 18). Between the two regression methods (linear multiple and random forest), similarities in the location of error clustering types by LISA and Gi* are observed. As a reminder, for LISA, the distinction is made between a statistically significant cluster consisting of high values only (HH), a cluster of low values only (LL), a cluster in which a high value is surrounded mainly by low values (HL), and a cluster in which a low value is surrounded mainly by high values (LH).

Figure 17. LISA of the differences between modelled and measured air temperatures for all study days (source: Data Grand Lyon).

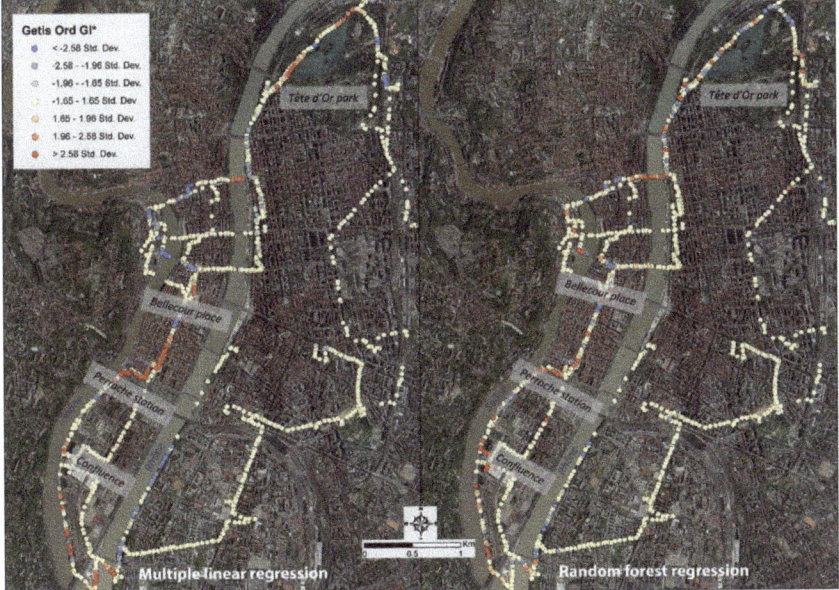

Figure 18. Gi* of the differences between modelled and measured air temperatures for all study days (source: Data Grand Lyon).

Firstly, using the LISA method, clusters of small errors (LH), underestimation of the model in relation to the measured values, can be identified on the left bank of the Rhône, in the steep streets of the peninsula and Vieux Lyon district, and on bridges. Areas with a high value (HL), i.e., an overestimation of the model, can be observed near the Perrache train station, in the Confluence area, and near the green spaces of the Tête d'Or park (Figure 17).

Secondly, groupings of the errors of underestimation and overestimation of the air temperature modelling compared to that measured by the Gi* method are located in areas similar to the LISA. These recurring areas for statistically low negative z-score values are the steep streets of the peninsula and Old Lyon and the south of the left bank of the Rhône. The statistically high positive z-score values are the Perrache train station and Confluence district, the proximity to the green spaces of the Tête d'Or park and the Morand bridge (Figure 18).

4.4. Limits and Future Research Outlooks

When looking at the positive or negative effects of variables on air temperature, it can be noticed that some can vary depending on the day being studied. For example, on 22 July 2019, the density of ground affects the air temperature positively, but on 19 July 2018, it had a negative impact. This is probably related to the route that differs between the two rides. The 2018 route is almost twice as long as the 2019 route (Figure 4) and the 2019 route passes through different neighborhoods, especially with regard to soil characteristics. This would indicate, among other reasons, why the results may differ depending on the days studied. In addition, the data provided by LiDAR concerning the ground is of a different nature, such as impermeable concrete or sandy soil for example, and may fluctuate depending on the routes taken. The same observation can be made for the proximity to metro stations. The proximity of the subway entrances is a variable that can affect air temperature in opposite ways. In our own experience, some subway entrances seem to give off fresh air and other entrances seem to give off warm air. When looking at the overall results of multiple linear regression modelling, it should be noted that these two variables are not included in the explanation of air temperature for these reasons.

The number of days processed for this study is one of its limitations. Indeed, only four days were analyzed. This limited number was partly due to the availability of quality (cloud-free) data from the Landsat satellite, but also due to the reduced occurrence of similar days in terms of climatic conditions. Another point of constraint is that modelling only took place in dense urban centers.

Consequently, we can argue on two perspectives: the spatial and temporal scope of this study. In the first case, it would be interesting to extend the mobile measurements in the periphery, or even in the rural areas, to be able to model the temperature in any point of the territory and compare the urban and the outskirts results. Secondly, it would be necessary to extend this analysis not only in summer, but in all seasons and at different moments of the day and at night, and for different weathers. Therefore, a global model could be built on observations from all the experimental dates rather than separating models by date.

In addition, some other data satellites may be used. For example, the use of the Sentinel 2 satellite with a 10 m resolution may help to increase the model results using sharper spectral indices, like NDVI or NDBaI.

5. Conclusions

The objective of this study was to identify the most appropriate and efficient regression to model urban air temperature based on numerous explanatory variables of various natures. The integration of these predictors in multiple regressions and machine learning method showed very satisfactory results. In addition, this methodology can be applied in other study area. The proportion of the variance explained by multiple linear regressions in air temperature modeling for each study day is globally high, with coefficients of determination ranging from 0.60 to 0.89. The results are even better when the random forest method is used. Indeed, the average coefficient of determination is 0.95 for a RMSE of only 0.17 °C and an OOB of 0.05. On the opposite, the PLS regression provides a weaker coefficient of determination for the separate days.

For all these models, there are recurring dominant variables such as NDVI or surface temperature. Consequently, the integration of satellite predictors is a definite advantage in urban microclimate modelling by linear regression model based on mobile air temperature measurements. In this study, Landsat 8 data were used, but one prospect for improvement would be to use higher resolution Sentinel data.

When we look at the overall results for all days combined, the same trends emerge. The multiple linear regression always gives very satisfactory results with an R^2 of 0.65 and an RMSE of 1.54 °C, on a par with the PLS regression which shows an R^2 of 0.70 and an RMSE of 1.50 °C. The global random forest modelling based on all days, however, proposes superior results with a high R^2 of 0.98 and an RMSE of 0.33 °C. This modelling method is therefore the most efficient of the three tested for this study area and this sample of measurements. However, it is less accessible than the other types of multiple regressions tested and requires a greater statistical investment.

One of the strengths of this study is also the fact that it is relatively easily applicable to other areas. The equipment used for mobile measurements is not very expensive. All that is needed is a radiation shelter, a GPS, and a temperature and relative humidity recorder. All the explanatory variables used in this study, such as land use area or satellite data, are freely available. GIS and statistical processing can also be freely available if one wishes to dispense with paying software. From a practical point of view, the most complicated part of the study remains the mobile field measurements, which are very time-consuming. Indeed, they have to be synchronized with the passage of Landsat and it is necessary to have similar and favorable weather conditions, with a completely clear sky and no wind.

The results of this study confirmed the cooling roles played by green areas and water surfaces and the problems linked to building density without vegetation in the urban overheating issue. In addition, low vegetation displayed low cooling power, mainly because of an absence of shade compared to the high vegetation and the low-density vegetation providing little evapotranspiration. This highlights the real need to use green and blue spaces solutions in order to limit the UHI and improve the thermal comfort.

Author Contributions: Conceptualization, L.A. and F.R.; methodology, L.A. and F.R.; validation, L.A. and F.R.; formal analysis, L.A. and F.R.; writing—original draft preparation, L.A. and F.R.; writing—review and editing, L.A. and F.R.; supervision, L.A. and F.R.; and project administration, L.A. and F.R. All authors have read and agreed to the published version of the manuscript.

Funding: This research received no external funding.

Acknowledgments: The authors gratefully acknowledge the EROS USGS, the Lyon Metropolis and the other data platform for the useful data, free of charge. This work would not have been possible without them.

Conflicts of Interest: The authors declare no conflict of interest.

Appendix A. Explanatory Variables Selected to Estimate Fine-Scale Air Temperature

Data Category	Variables Used for the Input (Units)	Expected Effect of the Variable on the Model	Calculation Method	Reference
Climatic data from remote sensing	Surface temperature (°C)	Positive	Single channel algorithm	[49,89,121,122]
	UTFVI Urban Thermal Field Variation Index)	Positive	$UTFVI = \frac{T_s - T_{main}}{T_s}$	[87,123]
	Brightness temperatures (°C)	Positive	$Brightness = \frac{K_2}{Ln\left(\frac{K_1}{L_s}+1\right)}$	[124,125]

Data Category	Variables Used for the Input (Units)	Expected Effect of the Variable on the Model	Calculation Method	Reference
Vegetation index	NDVI Normalized Difference Vegetation Index	Negative	$NDVI = \frac{NIR-RED}{NIR+RED}$	[85,126,127]
	SAVI Soil Adjusted Vegetation Index	Negative	$SAVI = \frac{NIR-RED}{NIR+RED+L} \times (L+1)$	[126]
	EVI Enhanced Vegetation Index	Negative	$EVI = G \times \frac{NIR-RED}{NIR+C_1 \times RED - C_2 \times BLUE + L}$	[126]
	Tasseled Cap greenness or GVI	Negative	$TCT\ G =$ Blue band $\times coefGr$ + Green band $\times coefGr$ + Red band $\times coefGr$ + NearInfrared band $\times coefGr$ + SWIR1 band $\times coefGr$ + SWIR2 band $\times coefGr$	[128]
	Density of low vegetation	Positive	LasTool Software (LasTool: http://lastools.org/) Vegetation quantity according to different buffer size	[46,97]
	Density of medium vegetation	Negative	LasTool Software Vegetation quantity according to different buffer size	[46,97]
	Density of high vegetation	Negative	LasTool Software Vegetation quantity according to different buffer size	[110]
Water presence index	NDWI Normalized Difference Water Index	Negative	$NDWI = \frac{Green-NIR}{Green+NIR}$	[85,126]
	MNDWI Modified Normalized Difference Water Index	Negative	$MNDWI = \frac{Green-SWIR1}{Green+SWIR1}$	[126]
Moisture index	Tasseled Cap Wetness	Negative	$TCT\ W =$ Blue band $\times coefWr$ + Green band $\times coefWr$ + Red band $\times coefWr$ + NearInfrared band $\times coefWr$ + SWIR1 band $\times coefWr$ + SWIR2 band $\times coefWr$	[128]
	NDMI Normalized Difference Moisture Index	Negative	$NDMI = \frac{NIR-SWIR1}{NIR+SWIR1}$	[86,88]
Bare soil index	NDBaI Normalized Difference Bareness Index	Positive	$NDBaI = \frac{SWIR1-TIRS}{SWIR1-TIRS}$	[85,126]
	BI Bare Soil Index	Positive	$BI = \frac{(SWIR1+RED)-(NIR+BLUE)}{(SWIR1+RED)+(NIR+BLUE)}$	[126]
	EBBI Enhanced Built-Up and Bareness Index	Positive	$EBBI = \frac{SWIR1-NIR}{10\sqrt{SWIR1+TIRS1}}$	[126]

Data Category	Variables Used for the Input (Units)	Expected Effect of the Variable on the Model	Calculation Method	Reference
Building index	NDBI Normalized Difference Built-Up Index	Positive	$NDBI = \frac{SWIR1-NIR}{SWIR1+NIR}$	[85,126]
	UI Urban Index	Positive	$UI = \frac{SWIR2-NIR}{SWIR2+NIR}$	[126]
	IBI Index-based Built-Up Index	Positive	$IBI = \frac{NDBI - \frac{(SAVI+MNDWI)}{2}}{NDBI + \frac{SAVI+MNDWI}{2}}$	[126]
	Building density	Positive	LasTool Software Building quantity according to different buffer size	[46,97]
Topographic	Slope (%)	Depending on the context	From the DEM (RVT 1.3 Software (RVT 1.3: https://iaps.zrc-sazu.si/en/rvt#v))	[129,130]
	Exposure (°N)	Depending on the context	From the DEM (RVT 1.3 Software)	[131]
	Curvature	Depending on the context	From the DEM (RVT 1.3 Software)	[132,133]
Proximity to land occupations	Water area	Negative	Water area according to different buffer size	[134,135]
	Distance to fountains	Negative	Euclidean distance to nearest fountain	
	Distance to subway entrances	Depending on the context	Euclidean distance to the nearest subway entrance	
	Distance to points of tourist interest	Negative	Euclidean distance to the nearest tourist point	
	Distance to railway tracks	Positive	Length of the railways according to different buffer size	
Radiation index	Spectral Radiance	Negative	$L_\lambda = L_{min(\lambda)} + \left(L_{max(\lambda)} - L_{min(\lambda)}\right)\frac{Q_{dn}}{Q_{max}}$	[136]
	Emissivity	Negative	$\epsilon = \frac{L_T}{L_{nT}}$	[137]
	Tasseled Cap Brightness	Positive	$TCT\ B = Blue\ band \times coefBr$ $+Green\ band$ $\times coefBr + Red\ band$ $\times coefBr$ $+ NearInfraredband$ $\times coefBr$ $+ SWIR1\ band$ $\times coefBr$ $+ SWIR2\ band$ $\times coefBr$	[128]
Urban morphology	Sky View Factor	Depending on the context	RVT 1.3 Software	[16,111,138]
	Variation in building height	Negative	Standard deviation of the building height	[97,116,139]

References

1. GIEC. Rapport Spécial du GIEC Sur le Réchauffement Planétaire de 1.5 °C. Organisation Météorologique Mondiale. Available online: https://public.wmo.int/fr/ressources/bulletin/rapport-sp%C3%A9cial-du-giec-sur-le-r%C3%A9chauffement-plan%C3%A9taire-de-15-%C2%B0c (accessed on 6 January 2020).
2. Oke, T.R. Canyon geometry and the nocturnal urban heat island: Comparison of scale model and field observations. *J. Clim.* **1981**, *1*, 237–254. [CrossRef]
3. Weston, K.J. Boundary layer climates. *Q. J. R. Meteorol. Soc.* **1988**, *114*, 1568. [CrossRef]
4. Oke, T. City size and the urban heat island. *Atmos. Environ.* **1973**, *7*, 769–779. [CrossRef]
5. Comtois, P.; Oke, T.R. Boundary layer climates, londres, methuen, 372 p., 15, 5 × 23, 5 cm, $18, 95. *Géo. Phys. Quat.* **1978**, *32*, 290–291. [CrossRef]

6. Aram, F.; García, E.H.; Solgi, E.; Mansournia, S. Urban green space cooling effect in cities. *Heliyon* **2019**, *5*, e01339. [CrossRef]
7. Buyadi, S.N.A.; Naim, W.; Mohd, W.; Misni, A. Quantifying green space cooling effects on the urban microclimate using remote sensing and GIS techniques. In Proceedings of the XXV International Federation of Surveyors, Kuala Lumpur, Malaysia, 16–21 June 2014.
8. Lahme, E.; Bruse, M. *Microclimatic Effects of a Small Urban Park in Densely Built-up Areas: Measurements and Model Simulations*; ICUC5: Lodz, Poland, 2003; Volume 5.
9. Oliveira, S.; Andrade, H.; Vaz, T. The cooling effect of green spaces as a contribution to the mitigation of urban heat: A case study in Lisbon. *Build. Environ.* **2011**, *46*, 2186–2194. [CrossRef]
10. Robitu, M.; Musy, M.; Inard, C.; Groleau, D. Modeling the influence of vegetation and water pond on urban microclimate. *Sol. Energy* **2006**, *80*, 435–447. [CrossRef]
11. Slater, G. *The Cooling Ability of Urban Parks*; University of Guelph: Guelph, ON, Canada, 2010. Available online: https://www.asla.org/2010studentawards/169.html (accessed on 16 July 2020).
12. Spronken-Smith, R.A.; Oke, T.R.; Lowry, W.P. Advection and the surface energy balance across an irrigated urban park. *Int. J. Climatol.* **2000**, *20*, 1033–1047. [CrossRef]
13. Yang, H.; Yang, K.; Miao, Y.; Wang, L.; Ye, C. Comparison of potential contribution of typical pavement materials to heat island effect. *Sustainability* **2020**, *12*, 4752. [CrossRef]
14. Erell, E.; Williamson, T. Intra-urban differences in canopy layer air temperature at a mid-latitude city. *Int. J. Clim.* **2007**, *27*, 1243–1255. [CrossRef]
15. Hodul, M.; Knudby, A.; Ho, H.C. Estimation of continuous urban sky view factor from landsat data using shadow detection. *Remote Sens.* **2016**, *8*, 568. [CrossRef]
16. Svensson, M.K. Sky view factor analysis—Implications for urban air temperature differences. *Meteorol. Appl.* **2004**, *11*, 201–211. [CrossRef]
17. Meng, C.; Dou, Y. Quantifying the anthropogenic footprint in Eastern China. *Sci. Rep.* **2016**, *6*, 24337. [CrossRef] [PubMed]
18. Yang, J.; Santamouris, M. Urban heat island and mitigation technologies in Asian and Australian Cities—impact and mitigation. *Urban Sci.* **2018**, *2*, 74. [CrossRef]
19. Zhou, D.; Zhao, S.; Zhang, L.; Sun, G.; Liu, Y. The footprint of urban heat island effect in China. *Sci. Rep.* **2015**, *5*, 11160. [CrossRef]
20. Ma, S.; Pitman, A.; Hart, M.; Evans, J.; Haghdadi, N.; MacGill, I. The impact of an urban canopy and anthropogenic heat fluxes on Sydney's climate. *Int. J. Clim.* **2017**, *37*, 255–270. [CrossRef]
21. United Nations. Climate Change Summit. Available online: https://www.un.org/en/climatechange/cities-pollution.shtml (accessed on 24 June 2020).
22. Alonso, L.; Renard, F. A Comparative study of the physiological and socio-economic vulnerabilities to heat waves of the population of the metropolis of Lyon (France) in a climate change context. *Int. J. Environ. Res. Public Heal.* **2020**, *17*, 1004. [CrossRef]
23. Dousset, B.; Gourmelon, F.; Laaidi, K.; Zeghnoun, A.; Giraudet, E.; Bretin, P.; Mauri, E.; Vandentorre, S. Satellite monitoring of summer heat waves in the Paris metropolitan area. *Int. J. Clim.* **2010**, *31*, 313–323. [CrossRef]
24. Dousset, B.; Gourmelon, F. Satellite multi-sensor data analysis of urban surface temperatures and landcover. *ISPRS J. Photogramm. Remote Sens.* **2003**, *58*, 43–54. [CrossRef]
25. Meehl, G.A. More intense, more frequent, and longer lasting heat waves in the 21st Century. *Science* **2004**, *305*, 994–997. [CrossRef]
26. Fallmann, J.; Forkel, R.; Emeis, S. Secondary effects of urban heat island mitigation measures on air quality. *Atmos. Environ.* **2016**, *125*, 199–211. [CrossRef]
27. Benas, N.; Chrysoulakis, N.; Cartalis, C. Trends of urban surface temperature and heat island characteristics in the Mediterranean. *Theor. Appl. Clim.* **2016**, *130*, 807–816. [CrossRef]
28. Chapman, S.; Watson, J.; Salazar, Á.; Thatcher, M.; McAlpine, C.A. The impact of urbanization and climate change on urban temperatures: A systematic review. *Landsc. Ecol.* **2017**, *32*, 1921–1935. [CrossRef]
29. Kukla, G.; Gavin, J.; Karl, T.R. Urban Warming. *J. Clim. Appl. Meteorol.* **1986**, *25*, 1265–1270. [CrossRef]
30. Sun, Y.; Gao, C.; Li, J.; Wang, R.; Liu, J. Quantifying the effects of urban form on land surface temperature in subtropical high-density urban areas using machine learning. *Remote Sens.* **2019**, *11*, 959. [CrossRef]

31. Wang, Y.; Ji, W.; Yu, X.; Xu, X.; Jiang, N.; Wang, Z.; Zhuang, D.-F. The impact of urbanization on the annual average temperature of the past 60 years in Beijing. *Adv. Meteorol.* **2014**, *2014*, 374987. [CrossRef]
32. Bobb, J.F.; Peng, R.D.; Bell, M.L.; Dominici, F. Heat-related mortality and adaptation to heat in the United States. *Environ. Heal. Perspect.* **2014**, *122*, 811–816. [CrossRef]
33. Renard, F.; Alonso, L.; Fitts, Y.; Hadjiosif, A.; Comby, J. Evaluation of the effect of urban redevelopment on surface Urban Heat Islands. *Remote Sens.* **2019**, *11*, 299. [CrossRef]
34. Rosenfeld, A.; Romm, J.; Akbari, H.; Pomerantz, M.; Taha, H. *Policies to Reduce Heat Islands: Magnitudes of Benefits and Incentives to Achieve Them*; Lawrence Berkeley National Lab.: Berkeley, CA, USA, 1996.
35. Akbari, H.; Pomerantz, M.; Taha, H. Cool surfaces and shade trees to reduce energy use and improve air quality in urban areas. *Sol. Energy* **2001**, *70*, 295–310. [CrossRef]
36. Santamouris, M.; Cartalis, C.; Synnefa, A.; Kolokotsa, D. On the impact of urban heat island and global warming on the power demand and electricity consumption of buildings—A review. *Energy Build.* **2015**, *98*, 119–124. [CrossRef]
37. Santamouris, M. On the energy impact of urban heat island and global warming on buildings. *Energy Build.* **2014**, *82*, 100–113. [CrossRef]
38. Chen, M.; Ban-Weiss, G.A.; Sanders, K.T. The role of household level electricity data in improving estimates of the impacts of climate on building electricity use. *Energy Build.* **2018**, *180*, 146–158. [CrossRef]
39. McPherson, E.G.; Nowak, D.; Heisler, G.; Grimmond, C.S.B.; Souch, C.; Grant, R.; Rowntree, R. Quantifying urban forest structure, function, and value: The Chicago Urban Forest Climate Project. *Urban Ecosyst.* **1997**, *1*, 49–61. [CrossRef]
40. Cristóbal, J.; Ninyerola, M.; Pons, X. Modeling air temperature through a combination of remote sensing and GIS data. *J. Geophys. Res. Space Phys.* **2008**, *113*. [CrossRef]
41. Kustas, W.P.; Norman, J.M. Use of remote sensing for evapotranspiration monitoring over land surfaces. *Hydrol. Sci. J.* **1996**, *41*, 495–516. [CrossRef]
42. Alonso, L.; Renard, F. Integrating satellite-derived data as spatial predictors in multiple regression models to enhance the knowledge of air temperature patterns. *Urban Sci.* **2019**, *3*, 101. [CrossRef]
43. De Ridder, K.; Maiheu, B.; Lauwaet, D.; Daglis, I.A.; Keramitsoglou, I.; Kourtidis, K.; Manunta, P.; Paganini, M. Urban Heat Island Intensification during Hot Spells—The Case of Paris during the Summer of 2003. *Urban Sci.* **2017**, *1*, 3. [CrossRef]
44. Leconte, F.; Bouyer, J.; Claverie, R.; Pétrissans, M. Analysis of nocturnal air temperature in districts using mobile measurements and a cooling indicator. *Theor. Appl. Clim.* **2016**, *130*, 365–376. [CrossRef]
45. Journal officiel du Sénat. Fermetures de Stations Météo-France et Avenir du Service Public Météorologique Français—Sénat. Available online: https://www.senat.fr/questions/base/2011/qSEQ110317685.html (accessed on 25 April 2019).
46. Foissard, X.; Dubreuil, V.; Quénol, H. Defining scales of the land use effect to map the urban heat island in a mid-size European city: Rennes (France). *Urban Clim.* **2019**, *29*, 100490. [CrossRef]
47. Richard, Y.; Emery, J.; Dudek, J.; Pergaud, J.; Chateau-Smith, C.; Zito, S.; Rega, M.; Vairet, T.; Castel, T.; Thévenin, T.; et al. How relevant are local climate zones and urban climate zones for urban climate research? Dijon (France) as a case study. *Urban Clim.* **2018**, *26*, 258–274. [CrossRef]
48. Nichol, J.E.; To, P.H. Temporal characteristics of thermal satellite images for urban heat stress and heat island mapping. *ISPRS J. Photogramm. Remote Sens.* **2012**, *74*, 153–162. [CrossRef]
49. Tsin, P.K.; Knudby, A.; Krayenhoff, E.S.; Ho, H.C.; Brauer, M.; Henderson, S.B. Microscale mobile monitoring of urban air temperature. *Urban Clim.* **2016**, *18*, 58–72. [CrossRef]
50. Mira, M.; Ninyerola, M.; Batalla, M.; Pesquer, L.; Pons, X. Improving mean minimum and maximum month-to-month air temperature surfaces using satellite-derived land surface temperature. *Remote Sens.* **2017**, *9*, 1313. [CrossRef]
51. Boer, E.P.J.; de Beurs, K.M.; Hartkamp, A.D. Kriging and thin plate splines for mapping climate variables. *Int. J. Appl. Earth Obs. GeoInf.* **2001**, *3*, 146–154. [CrossRef]
52. Jarvis, C.H.; Stuart, N. A Comparison among strategies for interpolating maximum and minimum daily air temperatures. Part II: The interaction between number of guiding variables and the type of interpolation method. *J. Appl. Meteorol.* **2001**, *40*, 1075–1084. [CrossRef]
53. Antonić, O.; Križan, J.; Marki, A.; Bukovec, D. Spatio-temporal interpolation of climatic variables over large region of complex terrain using neural networks. *Ecol. Model.* **2001**, *138*, 255–263. [CrossRef]

54. Wang, M.; He, G.; Zhang, Z.; Wang, G.; Zhang, Z.; Cao, X.; Wu, Z.; Liu, X. Comparison of spatial interpolation and regression analysis models for an estimation of monthly near surface air temperature in China. *Remote Sens.* **2017**, *9*, 1278. [CrossRef]
55. Zhang, Z.; Du, Q. A bayesian kriging regression method to estimate air temperature using remote sensing data. *Remote Sens.* **2019**, *11*, 767. [CrossRef]
56. Chen, Y.; Quan, J.; Zhan, W.; Guo, Z. Enhanced statistical estimation of air temperature incorporating nighttime light data. *Remote Sens.* **2016**, *8*, 656. [CrossRef]
57. Cristóbal, J.; Ninyerola, M.; Pons, X.; Pla, M. Improving air temperature modelization by means of remote sensing variables. In Proceedings of the 2006 IEEE International Symposium on Geoscience and Remote Sensing, Denver, CO, USA, 31 July–4 August 2006; Institute of Electrical and Electronics Engineers (IEEE): Piscataway, NJ, USA, 2006; Volume 2006, pp. 2251–2254.
58. Hengl, T.; Heuvelink, G.B.; Tadić, M.P.; Pebesma, E. Spatio-temporal prediction of daily temperatures using time-series of MODIS LST images. *Theor. Appl. Clim.* **2011**, *107*, 265–277. [CrossRef]
59. Zhu, W.; Lű, A.; Jia, S. Estimation of daily maximum and minimum air temperature using MODIS land surface temperature products. *Remote Sens. Environ.* **2013**, *130*, 62–73. [CrossRef]
60. Lhotellier, R. Spatialisation de la température minimale de l'air à échelle quotidienne sur quatre départements alpins français. *Climatologie* **2006**, *3*, 55–69. [CrossRef]
61. Zhou, B.; Erell, E.; Hough, I.; Shtein, A.; Just, A.C.; Novack, V.; Rosenblatt, J.; Kloog, I. Estimation of hourly near surface air temperature across israel using an ensemble model. *Remote Sens.* **2020**, *12*, 1741. [CrossRef]
62. Van Der Schriek, T.; Varotsos, K.V.; Giannakopoulos, C.; Founda, D. projected future temporal trends of two different urban heat islands in Athens (Greece) under three climate change scenarios: A statistical approach. *Atmosphere* **2020**, *11*, 637. [CrossRef]
63. Voelkel, J.; Shandas, V.; Haggerty, B. Developing high-resolution descriptions of urban heat islands: A public health imperative. *Prev. Chronic Dis.* **2016**, *13*. [CrossRef] [PubMed]
64. Leconte, F. Caractérisation des Îlots de Chaleur Urbain par Zonage Climatique et Mesures Mobiles: Cas de Nancy. Available online: http://www.theses.fr/2014LORR0255 (accessed on 24 June 2020).
65. Taha, H.; Levinson, R.; Mohegh, A.; Gilbert, H.; Ban-Weiss, G.A.; Chen, S. Air-temperature response to neighborhood-scale variations in albedo and canopy cover in the real world: Fine-resolution meteorological modeling and mobile temperature observations in the los angeles climate archipelago. *Climate* **2018**, *6*, 53. [CrossRef]
66. Fung, W.Y.; Lam, K.S.; Nichol, J.; Wong, M.S. Derivation of nighttime urban air temperatures using a satellite thermal image. *J. Appl. Meteorol. Clim.* **2009**, *48*, 863–872. [CrossRef]
67. Dobrovolný, P.; Krahula, L. The spatial variability of air temperature and nocturnal urban heat island intensity in the city of Brno, Czech Republic. *Morav. Geogr. Rep.* **2015**, *23*, 8–16. [CrossRef]
68. Charfi, S. Le Comportement Spatio-Temporel de la Température Dans L'agglomération de Tunis. Available online: http://www.theses.fr/2012NICE2036 (accessed on 24 June 2020).
69. Charfi, S.; Dahech, S. Cartographie des Températures à Tunis par Modélisation Statistique et Télédétection. *Mappemonde* **2018**, *123*, 19–30. [CrossRef]
70. Dahech, S. Evolution de la répartition spatiale des températures de l'air et de surface dans l'agglomération de Sfax entre 1987 et Impact sur la consommation d'énergie en été. (Evolution of the spatial distribution of air and surface temperatures in the agglomeration of Sfax between 1987 and 2010. Impact on energy consumption in summer). *Climatologie* **2012**, 11–32. [CrossRef]
71. Heusinkveld, B.G.; Van Hove, L.W.A.; Jacobs, C.M.J.; Steeneveld, G.J.; Elbers, J.A.; Moors, E.J.; Holtslag, A.A.M. Use of a mobile platform for assessing urban heat stress in Rotterdam. In Proceedings of the 7th Conference on Biometeorology, Freiburg, Germany, 12–14 April 2010; pp. 433–438.
72. Liu, L.; Lin, Y.; Wang, D.; Liu, J. An improved temporal correction method for mobile measurement of outdoor thermal climates. *Theor. Appl. Clim.* **2016**, *129*, 201–212. [CrossRef]
73. Rajkovich, N.; Larsen, L. A Bicycle-based field measurement system for the study of thermal exposure in Cuyahoga County, Ohio, USA. *Int. J. Environ. Res. Public Heal.* **2016**, *13*, 159. [CrossRef] [PubMed]
74. Brandsma, T.; Wolters, D. Measurement and statistical modeling of the Urban Heat Island of the City of Utrecht (the Netherlands). *J. Appl. Meteorol. Clim.* **2012**, *51*, 1046–1060. [CrossRef]
75. Joly, D.; Nilsen, L.; Fury, R.; Elvebakk, A.; Brossard, T. Temperature interpolation at a large scale: Test on a small area in Svalbard. *Int. J. Clim.* **2003**, *23*, 1637–1654. [CrossRef]

76. Beltrando, G. La climatologie: Une science géographique. *Geo. Inf.* **2000**, *64*, 241–261. [CrossRef]
77. Siewert, J.; Kroszczynski, K. GIS data as a valuable source of information for increasing resolution of the WRF Model for Warsaw. *Remote Sens.* **2020**, *12*, 1881. [CrossRef]
78. Wu, X.; Xu, Y.; Chen, H. Study on the spatial pattern of an extreme heat event by remote sensing: A case study of the 2013 extreme heat event in the Yangtze River Delta, China. *Sustainability* **2020**, *12*, 4415. [CrossRef]
79. Köppen, W. Versuch einer klassifikation der klimate, vorzugsweise nach ihren beziehungen zur pflanzenwelt. (Schluss). (Attempt to classify the climates, preferably according to their relationship to the flora. (Enough). *Geogr. Z.* **1900**, *6*, 657–679.
80. Kottek, M.; Grieser, J.; Beck, C.; Rudolf, B.; Rubel, F. World Map of the Köppen-Geiger climate classification updated. *Meteorol. Z.* **2006**, *15*, 259–263. [CrossRef]
81. Liu, Y.; Peng, J.; Wang, Y. Efficiency of landscape metrics characterizing urban land surface temperature. *Landsc. Urban Plan.* **2018**, *180*, 36–53. [CrossRef]
82. Miaomiao, X.; Yanglin, W.; Meichen, F. An overview and perspective about causative factors of Surface Urban Heat Island Effects. *Prog. Geogr.* **2011**, *30*, 35–41. [CrossRef]
83. Chen, A.; Sun, R.; Chen, L.-D. Studies on urban heat island from a landscape pattern view: A review. *Acta Ecol. Sin.* **2012**, *32*, 4553–4565. [CrossRef]
84. Yang, Q.; Huang, X.; Li, J. Assessing the relationship between surface urban heat islands and landscape patterns across climatic zones in China. *Sci. Rep.* **2017**, *7*, 1–11. [CrossRef] [PubMed]
85. Chen, X.; Zhao, H.-M.; Li, P.-X.; Yin, Z.-Y. Remote sensing image-based analysis of the relationship between urban heat island and land use/cover changes. *Remote Sens. Environ.* **2006**, *104*, 133–146. [CrossRef]
86. Jin, S.; Sader, S.A. Comparison of time series tasseled cap wetness and the normalized difference moisture index in detecting forest disturbances. *Remote Sens. Environ.* **2005**, *94*, 364–372. [CrossRef]
87. Liu, L.; Zhang, Y. Urban Heat Island Analysis Using the Landsat TM Data and ASTER Data: A Case Study in Hong Kong. *Remote Sens.* **2011**, *3*, 1535–1552. [CrossRef]
88. Nguyen, A.K.; Liou, Y.-A.; Li, M.-H.; Tran, T.A. Zoning eco-environmental vulnerability for environmental management and protection. *Ecol. Indic.* **2016**, *69*, 100–117. [CrossRef]
89. Sobrino, J.A.; Jiménez-Muñoz, J.C.; Paolini, L. Land surface temperature retrieval from LANDSAT TM. *Remote Sens. Environ.* **2004**, *90*, 434–440. [CrossRef]
90. Guo, Y.-J.; Han, J.-J.; Zhao, X.; Dai, X.-Y.; Zhang, H. Understanding the role of optimized land use/land cover components in mitigating summertime intra-surface urban heat island effect: A study on downtown Shanghai, China. *Energies* **2020**, *13*, 1678. [CrossRef]
91. Di Paola, F.; Ricciardelli, E.; Cimini, D.; Cersosimo, A.; Di Paola, A.; Gallucci, D.; Gentile, S.; Geraldi, E.; LaRosa, S.; Nilo, S.T.; et al. MiRTaW: An algorithm for atmospheric temperature and water vapor profile estimation from ATMS measurements using a random forests technique. *Remote Sens.* **2018**, *10*, 1398. [CrossRef]
92. Sekulić, A.; Kilibarda, M.; Heuvelink, G.B.M.; Nikolić, M.; Bajat, B. Random forest spatial interpolation. *Remote Sens.* **2020**, *12*, 1687. [CrossRef]
93. Dempster, A.P. Upper and lower probability inferences for families of hypotheses with monotone density ratios. *Ann. Math. Stat.* **1969**, *40*, 953–969. [CrossRef]
94. Shapiro, S.S.; Wilk, M.B. An analysis of variance test for normality (complete samples). *Biometrika* **1965**, *52*, 591–611. [CrossRef]
95. Dormann, C.F.; Elith, J.; Bacher, S.; Buchmann, C.; Carl, G.; Carré, G.; Márquez, J.R.G.; Gruber, B.; Lafourcade, B.; Leitão, P.J.; et al. Collinearity: A review of methods to deal with it and a simulation study evaluating their performance. *Ecography* **2012**, *36*, 27–46. [CrossRef]
96. Joint Research Centre—European Commission. *Handbook on Constructing Composite Indicators: Methodology and User Guide*; OCDE: Paris, France, 2008.
97. Shandas, V.; Voelkel, J.; Williams, J.; Hoffman, J.S. Integrating satellite and ground measurements for predicting locations of extreme urban heat. *Climate* **2019**, *7*, 5. [CrossRef]
98. Dempster, A.P. *Elements of Continuous Multivariate Analysis*; Addison-Wesley Pub. Co.: Boston, MA, USA, 1969.
99. Wold, S.; Sjöström, M.; Andersson, P.M.; Linusson, A.; Edman, M.; Lundstedt, T.; Nordén, B.; Sandberg, M.; Uppgård, L.-L. Multivariate Design and Modelling in QSAR, Combinatorial Chemistry, and Bioinformatics. In *Molecular Modeling and Prediction of Bioactivity*; Gundertofte, K., Jørgensen, F.S., Eds.; Springer: Boston, MA, USA, 2000. [CrossRef]

100. Tenenhaus, M.; Pagès, J.; Ambroisine, L.; Guinot, C. PLS methodology to study relationships between hedonic judgements and product characteristics. *Food Qual. Prefer.* **2005**, *16*, 315–325. [CrossRef]
101. Breiman, L.; Friedman, J.H.; Olshen, R.; Stone, C.J. *Classification and Regression Trees*; CRC Press: Boca Raton, FL, USA, 1984.
102. Hastie, T.; Tibshirani, R.; Friedman, J. *The Elements of Statistical Learning: Data Mining, Inference, and Prediction*, 2nd ed.; Springer: Berlin/Heidelberg, Germany, 2009.
103. Breiman, L. Bagging predictors. *Mach. Learn.* **1996**, *24*, 123–140. [CrossRef]
104. Gislason, P.O.; Benediktsson, J.A.; Sveinsson, J.R. Random Forests for land cover classification. Pattern Recognition Letters. Available online: https://dl.acm.org/doi/abs/10.1016/j.patrec.2005.08.011 (accessed on 30 May 2020).
105. Reid, S.; Tibshirani, R.; Friedman, J. A study of error variance estimation in Lasso regression. *Stat. Sin.* **2016**, *26*, 35–67. [CrossRef]
106. Tibshirani, R. Regression shrinkage and selection via the lasso. *J. R. Stat. Soc. Ser. B Stat. Methodol.* **1996**, *58*, 267–288. [CrossRef]
107. Anselin, L. Local indicators of spatial association-LISA. *Geogr. Anal.* **2010**, *27*, 93–115. [CrossRef]
108. Getis, A.; Ord, J.K. The analysis of spatial association by use of distance statistics. *Geogr. Anal.* **2010**, *24*, 189–206. [CrossRef]
109. Getis, A.; Ord, J. A research agenda for geographic information science. In *Spatial Analysis and Modeling in a GIS Environment*; Robert, B., McMaster, E., Lynn, U., Eds.; CRC Press: Boca Raton, FL, USA, 1996. Available online: https://books.google.fr/books?hl=fr&lr=&id=k9x0B3V3op0C&oi=fnd&pg=PA157&ots=cOnYyDRjKL&sig=nW-5WZ7_04hBe-lbgv2MdwBABBM&redir_esc=y#v=onepage&q&f=false (accessed on 3 May 2019).
110. Zhao, Q.; Yang, J.; Wang, Z.; Wentz, E.A. Assessing the cooling benefits of tree shade by an outdoor urban physical scale model at Tempe, AZ. *Urban Sci.* **2018**, *2*, 4. [CrossRef]
111. Chen, L.; Ng, E.; An, X.; Ren, C.; Lee, M.; Wang, U.; He, Z. Sky view factor analysis of street canyons and its implications for daytime intra-urban air temperature differentials in high-rise, high-density urban areas of Hong Kong: A GIS-based simulation approach. *Int. J. Clim.* **2010**, *32*, 121–136. [CrossRef]
112. Lin, P.; Lau, S.S.-Y.; Qin, H.; Gou, Z. Effects of urban planning indicators on urban heat island: A case study of pocket parks in high-rise high-density environment. *Landsc. Urban Plan.* **2017**, *168*, 48–60. [CrossRef]
113. Browder, G.; Ozment, S.; Rehberger Besco, I.; Gartner, T.; Lange, G.-M. *Integrating Green and Gray: Creating Next Generation Infrastructure*; World Bank: Washington, DC, USA; World Resources Institute: Washington, DC, USA, 2019. Available online: https://openknowledge.worldbank.org/handle/10986/31430 (accessed on 30 May 2020).
114. Gunawardena, K.; Wells, M.; Kershaw, T. Utilising green and bluespace to mitigate urban heat island intensity. *Sci. Total. Environ.* **2017**, *584*, 1040–1055. [CrossRef] [PubMed]
115. Colaninno, N.; Morello, E. Modelling the impact of green solutions upon the urban heat island phenomenon by means of satellite data. *J. Physics Conf. Ser.* **2019**, *1343*, 012010. [CrossRef]
116. Makido, Y.; Hellman, D.; Shandas, V. Nature-based designs to mitigate urban heat: The efficacy of green infrastructure treatments in Portland, Oregon. *Atmosphere* **2019**, *10*, 282. [CrossRef]
117. Eliasson, I.; Svensson, M.K. Spatial air temperature variations and urban land use—a statistical approach. *Meteorol. Appl.* **2003**, *10*, 135–149. [CrossRef]
118. Suomi, J.; Käyhkö, J. The impact of environmental factors on urban temperature variability in the coastal city of Turku, SW Finland. *Int. J. Clim.* **2011**, *32*, 451–463. [CrossRef]
119. Zhao, C.; Fu, G.; Liu, X.; Fu, F. Urban planning indicators, morphology and climate indicators: A case study for a north-south transect of Beijing, China. *Build. Environ.* **2011**, *46*, 1174–1183. [CrossRef]
120. Voogt, J.A.; Oke, T. Thermal remote sensing of urban climates. *Remote Sens. Environ.* **2003**, *86*, 370–384. [CrossRef]
121. Tran, H.; Uchihama, D.; Ochi, S.; Yasuoka, Y. Assessment with satellite data of the urban heat island effects in Asian mega cities. *Int. J. Appl. Earth Obs. GeoInf.* **2006**, *8*, 34–48. [CrossRef]
122. Gallo, K.; Hale, R.; Tarpley, D.; Yu, Y. Evaluation of the Relationship between Air and Land Surface Temperature under Clear- and Cloudy-Sky Conditions. *J. Appl. Meteorol. Clim.* **2011**, *50*, 767–775. [CrossRef]
123. Alfraihat, R.; Mulugeta, G.; Gala, T.S. Ecological evaluation of Urban Heat Island in Chicago City, USA. *J. Atmos. Pollut.* **2016**, *4*, 23–29. [CrossRef]

124. Pelta, R.; Chudnovsky, A.A.; Schwartz, J. Spatio-temporal behavior of brightness temperature in Tel-Aviv and its application to air temperature monitoring. *Environ. Pollut.* **2016**, *208*, 153–160. [CrossRef]
125. Walawender, J.P.; Szymanowski, M.; Hajto, M.J.; Bokwa, A. Land surface temperature patterns in the urban agglomeration of Krakow (Poland) derived from landsat-7/ETM+ data. *Pure Appl. Geophys.* **2014**, *171*, 913–940. [CrossRef]
126. Hasanlou, M.; Mostofi, N. Investigating Urban Heat Island Effects and Relation Between Various Land Cover Indices in Tehran City Using Landsat 8 Imagery. In Proceedings of the 1st International Electronic Conference on Remote Sensing, Basel, Switzerland, 22 May 2015; p. 1.
127. Roșca, C.F.; Harpa, G.V.; Croitoru, A.-E.; Herbel, I.; Imbroane, A.M.; Burada, D.C. The impact of climatic and non-climatic factors on land surface temperature in southwestern Romania. *Theor. Appl. Clim.* **2016**, *130*, 775–790. [CrossRef]
128. Baig, M.H.A.; Zhang, L.; Shuai, T.; Tong, Q. Derivation of a tasselled cap transformation based on Landsat 8 at-satellite reflectance. *Remote Sens. Lett.* **2014**, *5*, 423–431. [CrossRef]
129. Kim, Y.-H.; Baik, J.-J. Daily maximum urban heat island intensity in large cities of Korea. *Theor. Appl. Clim.* **2004**, *79*, 151–164. [CrossRef]
130. Weng, Q.; Firozjaei, M.K.; Sedighi, A.; Kiavarz, M.; Alavipanah, S.K. Statistical analysis of surface urban heat island intensity variations: A case study of Babol City, Iran. *GISci. Remote Sens.* **2018**, *56*, 576–604. [CrossRef]
131. Ali-Toudert, F.; Mayer, H. Effects of asymmetry, galleries, overhanging façades and vegetation on thermal comfort in urban street canyons. *Sol. Energy* **2007**, *81*, 742–754. [CrossRef]
132. Hafner, J.; Kidder, S.Q. Urban Heat Island modeling in conjunction with satellite-derived surface/soil parameters. *J. Appl. Meteorol.* **1999**, *38*, 448–465. [CrossRef]
133. Weng, Q.; Quattrochi, D.A. Thermal remote sensing of urban areas: An introduction to the special issue. *Remote Sens. Environ.* **2006**, *104*, 119–122. [CrossRef]
134. Shohei, K.; Takeki, I.; Hideo, T. Relationship between Terra/ASTER land surface temperature and ground-observed air temperature. *Geogr. Rev. Jpn. Ser. B* **2016**, *88*, 38–44. [CrossRef]
135. Madelin, M.; Bigot, S.; Duché, S.; Rome, S. Intensité et délimitation de l'îlot de chaleur nocturne de surface sur l'agglomération parisienne. In *Archive ouverte en Sciences de l'Homme et de la Société*; HAL: Houston, TX, USA, 2017; Volume 9.
136. Iizawa, I.; Umetani, K.; Ito, A.; Yajima, A.; Ono, K.; Amemura, N.; Onishi, M.; Sakai, S. time evolution of an urban heat island from high-density observations in Kyoto City. *SOLA* **2016**, *12*, 51–54. [CrossRef]
137. Taha, H. Urban climates and heat islands: Albedo, evapotranspiration, and anthropogenic heat. *Energy Build.* **1997**, *25*, 99–103. [CrossRef]
138. Qaid, A.; Ossen, D.R.; Rasidi, M.H.; Bin Lamit, H. Effect of the position of the visible sky in determining the sky view factor on micrometeorological and human thermal comfort conditions in urban street canyons. *Theor. Appl. Clim.* **2017**, *131*, 1083–1100. [CrossRef]
139. Voelkel, J.; Shandas, V. Towards systematic prediction of urban heat islands: Grounding measurements, assessing modeling techniques. *Climte* **2017**, *5*, 41. [CrossRef]

© 2020 by the authors. Licensee MDPI, Basel, Switzerland. This article is an open access article distributed under the terms and conditions of the Creative Commons Attribution (CC BY) license (http://creativecommons.org/licenses/by/4.0/).

Article

Automatic Pattern Recognition of Tectonic Lineaments in Seafloor Morphology to Contribute in the Structural Analysis of Potentially Hydrocarbon-Rich Areas

Eleni Kokinou [1],* and Costas Panagiotakis [2,3]

1. Laboratory of Applied Geology and Hydrogeology, Department of Agriculture, Hellenic Mediterranean University, Stavromenos, 71004 Heraklion, Greece
2. Laboratory of Data Science, Multimedia and Modelling, Department of Management Science and Technology, Hellenic Mediterranean University, 72100 Agios Nikolaos, Greece; cpanag@hmu.gr
3. Institute of Computer Science, Foundation for Research and Technology-Hellas, 70013 Heraklion, Greece
* Correspondence: ekokinou@hmu.gr

Received: 24 April 2020; Accepted: 10 May 2020; Published: 12 May 2020

Abstract: This work presents novel pattern recognition techniques applied on bathymetric data from two large areas in Eastern Mediterranean. Our objectives are as follows: (a) to demonstrate the efficiency of this methodology, (b) to highlight the quick and accurate detection of both hydrocarbon related tectonic lineaments and salt structures affecting seafloor morphology, and (c) to reveal new structural data in areas poised for hydrocarbon exploration. In our work, we first apply a multiple filtering and sequential skeletonization scheme inspired by the hysterisis thresholding technique. In a second stage, we categorize each linear and curvilinear segment on the seafloor skeleton (medial axis) based on the strength of detection as well as the length, direction, and spatial distribution. Finally, we compare the seafloor skeleton with ground truth data. As shown in this paper, the automatic extraction of the bathymetric skeleton allows the interpretation of the most prominent seafloor morphological features. We focus on the competent tracing of tectonic lineaments, as well as the effective distinction between seafloor features associated with shallow evaporite movements and those related to intense tectonic activity. The proposed scheme has low computational demand and decreases the cost of the marine research because it facilitates the selection of targets prior to data acquisition.

Keywords: Eastern Mediterranean Sea; multiple filtering; skeletonization; structural interpretation

1. Introduction

Submarine geomorphology studies the landforms (i.e., relief) and processes (tectonic, sedimentary, oceanographic, and biological) in the submarine realm, many of which comprise renewable and non-renewable resources in many maritime countries. Resources of importance to such countries include unique ecosystems, fishery resources, freshwater, aggregates, minerals, ocean-driven energy, and hydrocarbons [1]. In fact, ocean basins around the world host an extraordinary number of landforms (mid-ocean ridges, seamount/knoll/guyots, volcanic islands, rift basins, trenches, abyssal plains) where such resources are usually located. In addition, continental shelves, comprising the underwater portion of the continent in relatively shallow waters, are rich in marine life, minerals, and hydrocarbons. Marine geomorphometry deals with the characterization of the seafloor using geomorphometric techniques in digital bathymetric data, that is, terrain attributes such as slope, aspect, curvature, roughness, feature extraction, and automatic classification [2,3]. The link between tectonics and the shape of the ocean floor, as addressed in this work [4–6], is of particular importance to such geomorphometric analyses. Tectonic activity is often responsible for the formation of submarine landforms. However, a certain attenuation of resulting tectonic landscapes is expected as sediments fill and drape the ocean floor [7]. The degree of attenuation relates to local sedimentation rates, the degree of geodynamic activity in a certain area (earthquakes, submarine volcanism, seafloor spreading, and so on), fluid and mass transfers, slope instability processes, and the activity of gravity (gliding) tectonics as trigger for complex folding and faulting. In addition, evaporite (salt)-bearing basins around the world, associated with either compressional or extensional tectonic regimes, are one of the most attractive areas for hydrocarbon exploration. The Mediterranean along with the North Caspian, Mexican, and East Siberian salt-bearing basins are considered the four super-giants that contain evaporite masses of about 1.5–2.5 mln. km^3 each [8].

The automatic extraction of geomorphological features [6,9–18] using the digital elevation model (DEM) belongs to the techniques of remote sensing (RS) data processing, which enhances knowledge in geology and geomorphology. Previously published work on the automatic extraction of geomorphological features used the connectivity tree (topographic change) [9], growing segmentation on DEMs and appropriate gradient-region growing criteria [10], the classification of continuous topography using taxonomic criteria (surface texture, slope gradient, and local convexity), and spectral analysis methods allowing the production of dominant wavelength maps [13]. Automatic detection of mountains, topographic highs, and volcanos is implemented from the analysis of DEMs, providing geomorphological features useful for annotation and classification tasks [15,17,18]. The VOLEI method is an unsupervised iterative method based on volume evolution of isocontours to detect topographic highs and to quantify the terrain's morphology using four terrain's quantitative properties (i.e., orientation, average slope, eccentricity, and shape complexity [18]). The automatic detection of tectonic lineaments is realized by applying image processing methods on DEMs and remote sensing images like principal component analysis, filtering, and Hough transform [19]. The authors of [20,21] developed a quantitative method to recognize geological faults (strike-slip, dip-slip, and oblique) based on the calculation of the horizontal and vertical curvature from land DEMs. In addition, in [16], the automatic enhancement and identification of the linear and curvilinear elements and corresponding to tectonic lineaments/geological faults is implemented via a filtering technique combined with a graph sampling approach.

In this paper, we use medium- and high-resolution bathymetric data from the Southern Cretan offshore and the Levantine Basin, to the east and southwest of the Eratosthenes Seamount (Figure 1), to show how multiple filtering and skeletonization (automatic line-drawing) can contribute to the following:

(a) marine geomorphology/geomorphometrics and further structural studies;
(b) the accurate tracing of tectonic lineaments on the seafloor in large marine areas;
(c) the detection of morphostructures on the seafloor related to shallow evaporite movement known as halokinesis.

In order to address the previous aims, bathymetric skeletons (medial axes from bathymetric data) are initially computed based on recently developed multiple filtering and skeletonization algorithms [6,16,18]. Then, the linear and curvilinear elements of the bathymetric skeleton are categorized based on criteria such as the strength of detection, their length, direction, and spatial distribution. Finally, we compare the automatically calculated bathymetric skeleton with ground truth data. Both case studies (the Southern Cretan offshore and the Levantine Basin, Figure 1) present bathymetric skeletons that are indicative of active tectonics. In addition, the bathymetric pattern in the Levantine Basin, southwest of the Eratosthenes Seamount, mostly presents morphological characteristics related to shallow salt movements (Figure 1).

Evaporites are water-soluble, layered, crystalline sedimentary rocks formed from surface or near surface brines in areas where high solar evaporation rates prevailed in the geological past [22]. Evaporation of seawater initially produces calcium carbonate ($CaCO_3$) followed by calcium sulfate in the form of gypsum ($CaSO_4 \cdot 2H_2O$) or anhydrite ($CaSO_4$). The following mineral to form is halite (NaCl), commonly known as rock salt. Halite presents a low density (2160 kg/m^3) and is mechanically weak, particularly under heavy overburden loading and high temperatures derived from its progressive burial. Thus, salt can flow like a visco-elastic fluid, even in cases where geologically strain rates are rapid [23]. Differential loading induced by gravitational forces, variable thermal conditions, or salt boundary displacement is the primary force inducing salt movement and usually results in vertical and lateral spreading or gliding along an inclined substratum [23]. Buried evaporite often manages to reach shallow depths below the seafloor, giving rise to specific morphological structures and further lineaments, as shown later in this work. Generally, lineaments are the surface expression of geological features such as fractures; faults; folds; geological contacts; rivers; streams; and other physiographic structures showing different origin, size, age, and depth [24–27]. They are important in recognizing geological structures on Earth's surface related to the control and distribution of natural and mineral resources, geohazards, geothermal energy, and earthquakes. Tectonic lineaments relate to tectonic processes such as faulting and folding and are used in the production of structural maps.

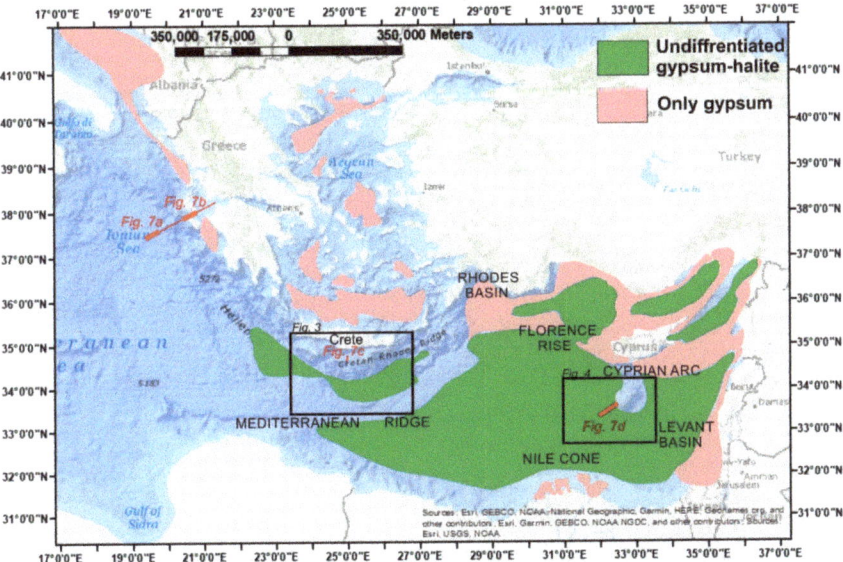

Figure 1. Map of the wide area of Eastern Mediterranean Sea and corresponding distribution of Late Miocene (Messinian) salt. The data concerning the base map are from the ArcGIS 10 online database. Black rectangles correspond to the automatic detection of bathymetric structures. The red lines in this figure show the locations of the seismic reflection profiles shown later in this work where the seafloor is deformed owing to the presence of underlying salt. Recently, updated maps concerning the distribution of both Triassic and Messinian salts in the Mediterranean Sea have been presented in previous works [8].

2. Materials and Methods

Medium- and high-resolution bathymetric data, from the open-file database EMODNet (European Marine Observatory and Data Network [28]) or compiled from various other sources [29–32], were primarily used in our research. Medium-resolution bathymetry derived from EMODNet comprises a grid/cell size of 0.25 arc minutes, while high-resolution bathymetry has a grid/cell size of 250 m.

The present methodology (Figure 2) continues the work of [16–18,33,34] that has been partially or in whole applied in previously published geological and environmental studies [6,35–40]. We summarize the steps of this methodology as follows:

2.1. Enhancement of the Seafloor Texture

Firstly, slope and aspect images as well their derivatives are computed. Then, the slope and aspect images are efficiently combined for the computation of the enhancement image F of seafloor texture (e.g., lineaments). The enhancement image F is computed as proposed in [16] and tested using onshore and offshore data [6,35–40].

$F(p) = (S^2(p) \times SS(p) \times SA(p))^{1/4}$ is calculated based on the slope $S(p)$, the aspect $A(p)$, the first derivative of the slope SS (p), and the first derivative of the aspect SA(p) at point p (S(p)) of the topographic surface.

2.2. Multiple Filtering

In the enhancement image F, a step filter $G(a,w)$ is applied, proposed in [34], to decrease noise and emphasize the linear and curvilinear structures of orientation a and width w.

$$I_g(a,w) = F(p) * G(a,w) \qquad (1)$$

The filter $G(a, w)$ was constrained to be zero mean. In addition, the filter energy is normalized to one under any orientation and width (filter parameters), so that the responses of different angles and widths are comparable, respectively.

In order to enhance the curvilinear structures under any orientation and width, the image I_m is computed by getting the maximum of the corresponding pixel values of $I_g(a, w)$.

2.3. Pixel Labeling

The preliminary goal of skeletonization is to classify I_m pixels into three classes, C_1, C_2, and C_3, with label numbers 1, 2, and 3, respectively:

C_1: The pixels corresponding to curvilinear structures.

C_2: The pixels uncertain to correspond to curvilinear structures. We decide for them in the last step, based on the other classes' classification (C_1 and C_3).

C_3: The pixels that do not correspond to curvilinear structures.

The definition of the above classes is based on the hysterisis thresholding technique, which has been used on edge detection problem [41]. According to hysterisis thresholding, two thresholds, T_l (low) and T_h (high), are used to initially classify the pixels into three classes (C_1, C_2, and C_3). The advantage of this thresholding type is the exclusion of some connected point groups [42]. In the proposed scheme, the thresholds T_l and T_h are automatically estimated based on the median value of I_m values (Med).

- T_l is given by the mean value of I_m values that reveal a value lower than Med.
- T_h is given by the mean value of I_m values that reveal a value higher than Med.

Let B_i be the image of initial pixel classification into classes C_1, C_2, and C_3 and $I_m(p)$ and $mv(p)$ to denote the value of image I_m on pixel p and the median value of the nine pixel-neighborhood of pixel p in I_m, respectively.

1. If $I_m(p) \geq T_h$ and $I_m(p) > mv(p)$, p is classified to C_1, as its value is very high compared with the image ($I_m(p) \geq T_h$) and with its neighbourhood ($I_m(p) > mv(p)$).
2. Else, if $I_m(p) \geq T_h$ or ($I_m(p) > T_l$ and $I_m(p) > mv(p)$), p is classified to C_2. This is true if the pixel value is high compared with the image, but it is not high enough compared with its neighbourhood, or reversely.
3. Otherwise, p is classified to C_3.

Finally, a region growing-based method is applied on pixels of C_2 class to provide the final pixel labeling into classes C_1 and C_3. Let B_f be the image of final pixel classification into classes C_1 and C_3.

According to the method, the pixels of C_2 class are classified to C_1 if they are connected to a pixel of C_1; otherwise, they are classified to C_3 class.

Image skeletonization (thin curvilinear structure detection) (B_t) is provided, if we change the second rule of classification to class C_2, removing the case of $I_m(p) \geq T_h$.

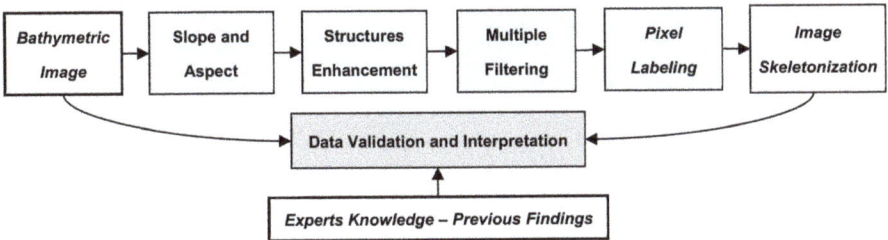

Figure 2. Flow-chart illustrating the methodology of the present work.

3. Results

3.1. Automatically Calculated Bathymetric Skeleton

Image skeletonization aims to enhance shape analysis and data interpretation. Figures 3 and 4 illustrate the results of the multiple filtering and skeletonization for both the Southern Cretan offshore and the area on, east, and southwest of the Eratosthenes Seamount, using the bathymetric data as input. The output of the multiple filtering approach for the Southern Cretan offshore and the area on, east, and southwest of the Eratosthenes Seamount is illustrated in Figures 3a and 4a, respectively. The strong and weak detections are depicted with yellow and blue colors, respectively. Finally, Figures 3b and 4b indicate the outcome of the thin curvilinear structure detection (skeletonization) using colour lines, projected on the original bathymetric relief. The red lines correspond to strongly detected linear and curvilinear elements, while the blue lines correspond to weak ones. The texture of the skeleton and changes in specific segments of the images are shown in Figures 3b and 4b. The differences in the skeleton texture possibly relate to (a) sedimentation rates, (b) the degree of geodynamic activity in a certain area, (c) fluid and mass transfers, (d) slope instability processes, and (e) the activity of gravity (gliding) tectonics. Because of the differences in the skeleton texture, we divided both study areas in sub-regions (A, B, C and A′, B′, C′), respectively, based on the following criteria:

- The strength of the detection using the color scale shown in Figures 3a and 4a ranging from 0 (blue) to 550 (yellow) with no units;
- The length of lineaments (short, medium, long);
- The direction of lineaments;
- The spatial distribution of the lineaments (sparse, medium, dense).

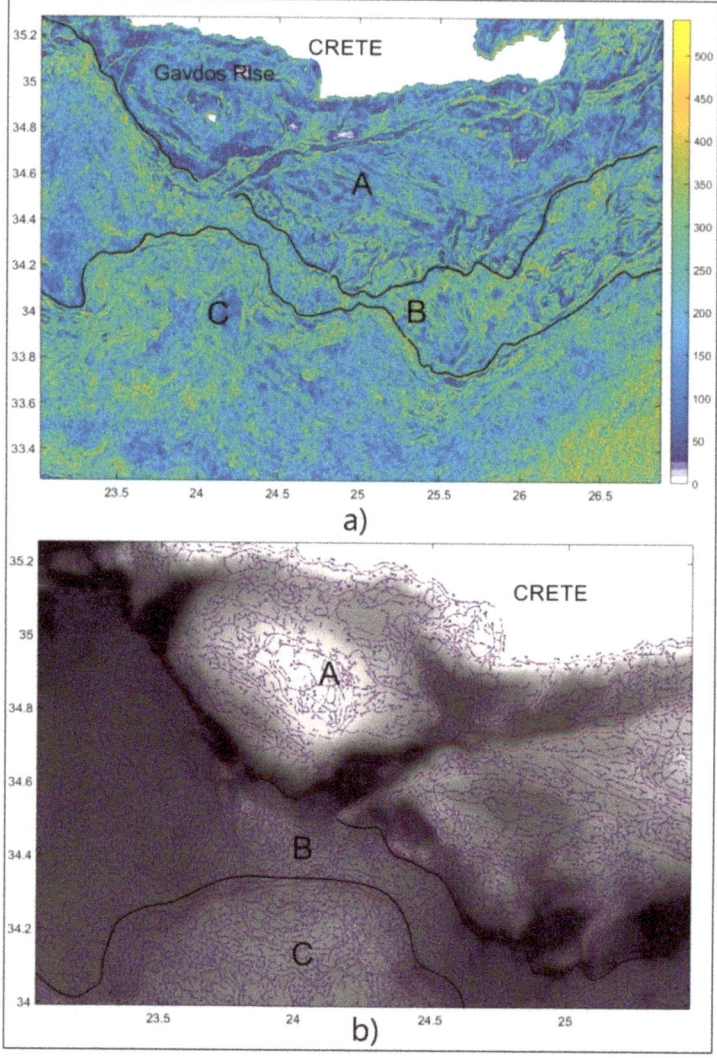

Figure 3. Automatic detection of the seafloor skeleton in the Southern Cretan offshore using medium resolution (0.25 arc minutes) data from EMODNet (European Marine Observatory and Data Network) [28]. The reference system is WGS84. (**a**) The output of the multiple filtering approach [34]. Color scale corresponds to the strength of the detection ranging from 0 (blue) to 550 (yellow) with no units. (**b**) The skeleton (medial axes) overlain the bathymetric relief. The red lines correspond to strongly detected linear and curvilinear elements, while the blue lines correspond to weak ones. On the basis of signal texture (strength of the detection, length, direction, and spatial distribution of the lineaments), we divided the area into three sub-regions (A, B, and C).

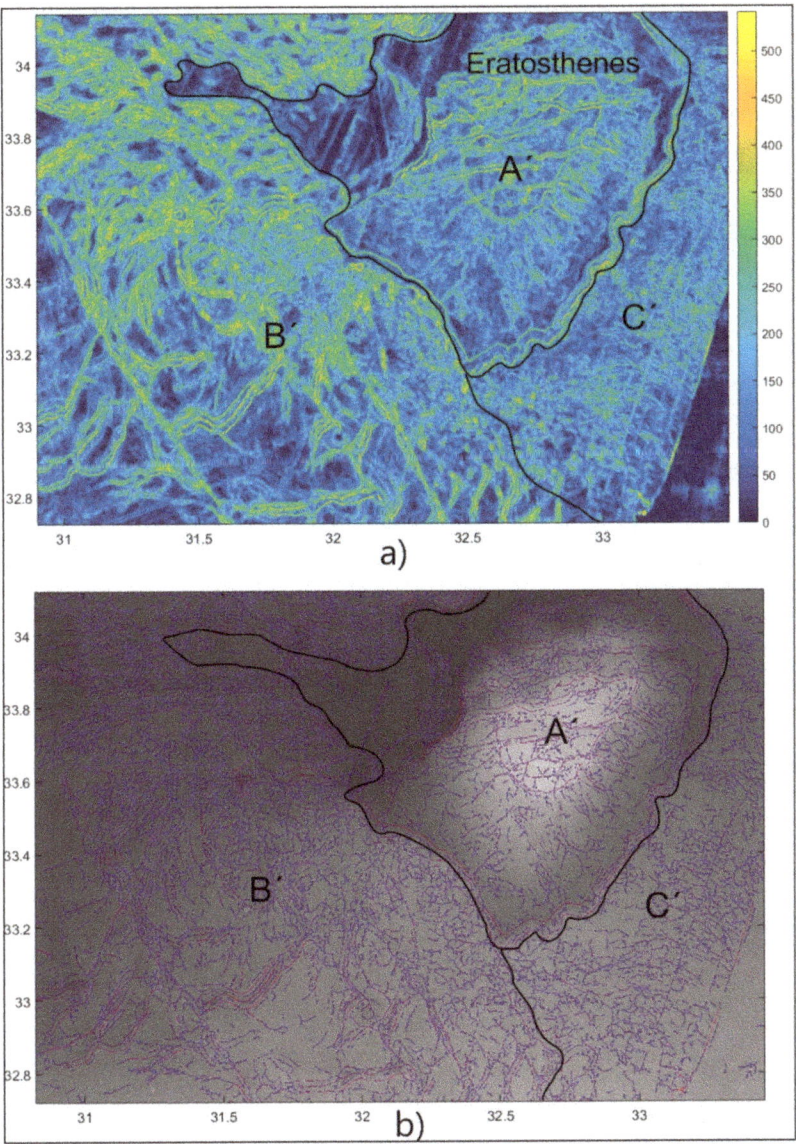

Figure 4. Automatic detection of the seafloor skeleton in the Levantine Basin (east and southwest of the Eratosthenes Seamount) using medium resolution (0.25 arc minutes) data from EMODNet (European Marine Observatory and Data Network) [28] and other sources [29–32]. The reference system is WGS84. (**a**) The output of the multiple filtering approach [34]. Color scale corresponds to the strength of the detection ranging from 0 (blue) to 550 (yellow) with no units. (**b**) The skeleton (medial axes) overlain the bathymetric relief. The red lines correspond to strongly detected linear and curvilinear elements, while the blue lines correspond to weak ones. On the basis of signal texture (strength of the detection, length, direction, and spatial distribution of the lineaments), we divided the area into three sub-regions (A', B', and C').

More specifically, the sub-regions A, B, C and A', B', C' (Figure 3a,b and Figure 4a,b) show the following characteristics taking into account the above criteria:

- Sub-region A: Located in the Southern Cretan offshore (Ptolemy and Pliny Trenches including Gavdos high), characterized by medium to long in length and medium to sparse distributed lineaments generally trending NE–SW, NW–SE, and E–W, which present high detection strength (Figure 3a,b).
- Sub-region B: Located in the Southern Cretan offshore (Strabo Trench and the region west of Gavdos high) characterized by short to medium in length and medium to dense distributed lineaments trending to all directions (Figure 3a,b). The lineaments in the southeastern part of sub-region B present high detection strength, while those in the southwestern part of sub-region B present low detection strength (Figure 3a).
- Sub-region C: Located in the Southern Cretan offshore (Mediterranean Ridge) characterized by medium to long in length and medium to sparse distributed lineaments. Lineaments, close to the limit with sub-regions B and C, present similar orientation with this limit (Figure 3a,b). In addition, the lineaments near to the limit with sub-regions B and C present high detection strength (Figure 3a). The lineaments away from the limit with sub-regions B and C are generally short in length, medium to sparse distributed, and show no prevailing orientation (Figure 3a,b).
- Sub-region A': Located on and west of the Eratosthenes Seamount characterized by (a) the presence of strong detected E–W trending, medium to long in length, and medium to sparse distributed lineaments on the Eratosthenes Seamount; and (b) very few, weak detected lineaments west of the Eratosthenes Seamount (Figure 4a,b).
- Sub-region B': Located southwest of the Eratosthenes Seamount characterized by (a) strong detected, NE–SW trending lineaments intersected by a group of NW–SE trending lineaments, both long in length and sparse distributed; and (b) medium to weak detected lineaments with no prevailing orientation that are short in length and sparse to medium distributed (Figure 4a,b).
- Sub-region C': Located east and southeast of the Eratosthenes Seamount characterized by generally weak detected, short in length and medium to sparse distributed lineaments that generally trend E–W to NW–SE (Figure 4a,b).

3.2. Comparison with Ground Truth Data

In Figure 5a,b, we present the comparison of the automatically calculated bathymetric skeleton with ground truth data for two selected areas. A significant part of the linear and curvilinear elements of the bathymetric skeleton is expected to correspond to tectonic lineaments, as both areas are strongly influenced by intense geodynamic activity related to different mechanisms.

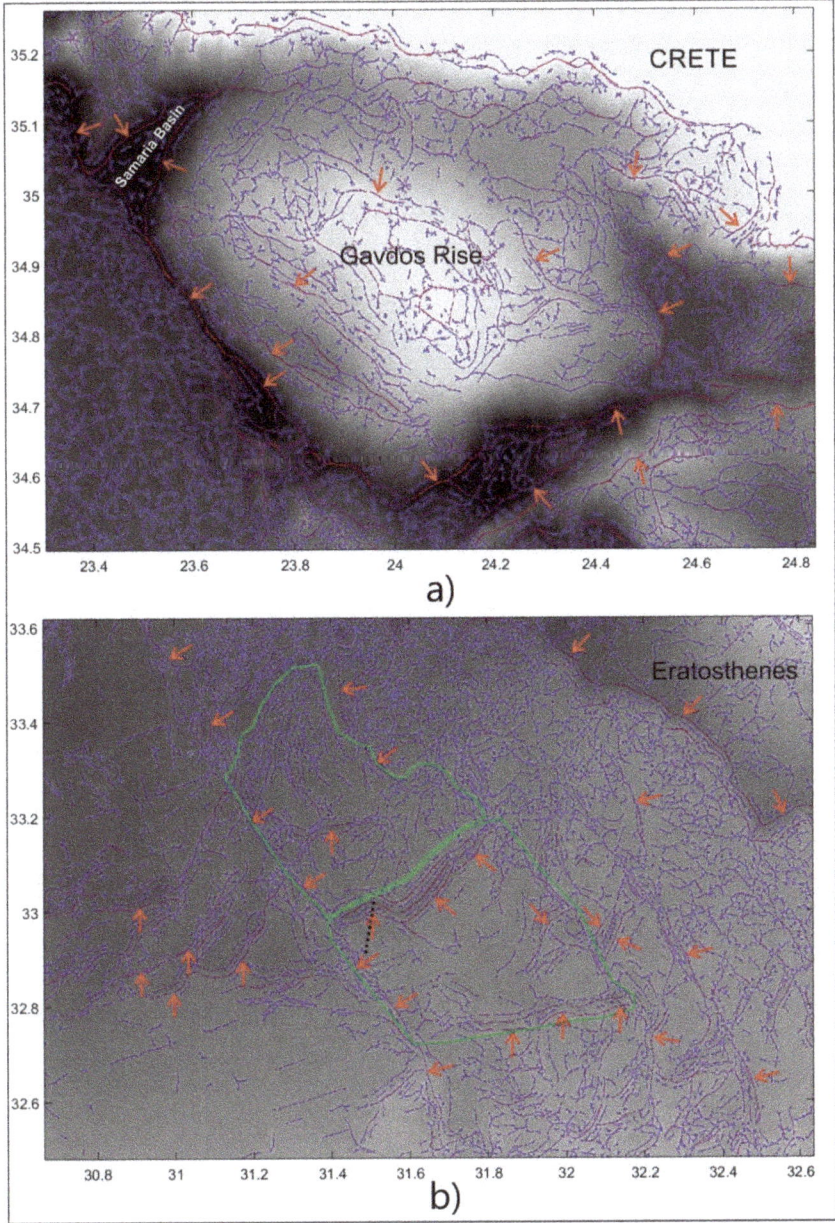

Figure 5. Comparison of the automatically detected linear and curvilinear elements with ground truth data. The red lines correspond to strongly detected linear and curvilinear elements, while the blue lines correspond to weak ones. Red arrows show the linear and curvilinear elements corresponding to strong automatic detection and they are in full agreement with previously reported geological faults in (**a**) the wide area of Gavdos Rise in South Crete [43–45] and (**b**) southwest of the Eratosthenes Seamount. Areas indicated by the closed green line are possibly affected by shallow salt movements [46–48].

The seafloor topography of the Southern Cretan Margin is strongly associated with transtensional movements in the upper 10–15 km of the crust, while transpression prevails below these depths [43]. Small-scale domes due to Messinian evaporitic intrusions may locally deform the recent stratigraphic successions (Pliocene–Quaternary) of the Southern Cretan Margin. Specifically, Figure 5a presents the automatically detected linear and curvilinear elements for the wide area of Gavdos Rise in South Crete. The red arrows show the most prominent automatically detected tectonic lineaments related to already reported geological faults [43–45]. Figure 6a,b present in more detail how we implemented the validation of the automatically detected tectonic lineaments. This figure resulted from the overlay of the geological faults (in red) [43–45] with the automatically detected skeleton. According to our previous work [16], the proposed method with linear patterns selection algorithm (LPSA) yields approximately 75% of the automatic detected lineaments to coincide either in location or in direction with the geological faults from ground truth data. The methodology in this work yields approximately 85% of the geological faults (thick red lines) to coincide in both location (less than 0.5 arc-minutes) and direction (less than 10 degrees) with the corresponding part of the bathymetric skeleton, meaning that the recall of the proposed method is about 85% (Figure 6b). Thus, there is a strong correlation between the automatically detected tectonic lineaments from the bathymetric skeleton (Figure 6) and those reported from previous studies [43–45]. In addition, a significant amount of information concerning the presence of additional tectonic lineaments is shown in Figure 5a, possibly related to primary or secondary fracture systems that should be investigated in the future.

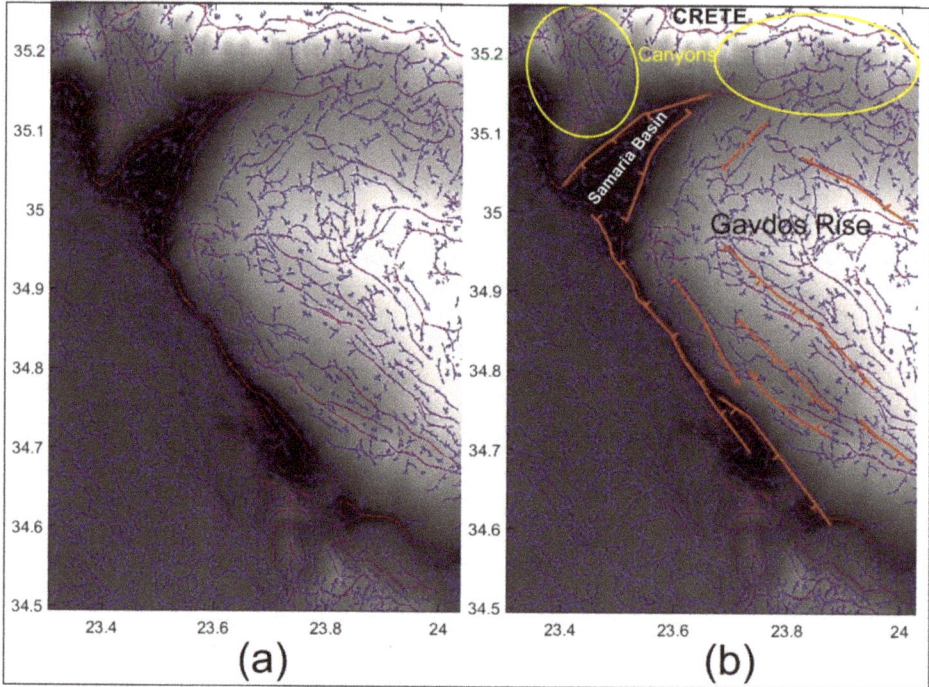

Figure 6. An example showing the comparison of the (**a**) automatically detected linear and curvilinear elements from the western part of Gavdos Rise in South Crete with (**b**) geological faults (thick red lines) from previous research [43–45] and references therein. The yellow ellipsis indicates areas of canyons [45].

The deformation of the area southwest of the Eratosthenes Seamount is mostly driven by salt tectonics related to the presence of underlying thick Messinian evaporites triggering processes of regional gravity spreading and gliding [46]. Specifically, the authors of [46], based on seismic and swath bathymetric data for this area, reported the presence of basins anchored within pre-Messinian sequences. This is because of massive salt removal and displacement of the salt/sediment system to areas with smoother slopes. Two examples of such areas, indicated by green line, are shown in Figure 5b. These areas (Figure 5b) are limited by distinctive, linear or curvilinear, continuous, strongly detected lineaments that probably correspond to large-scale geological faults [46,47]. In addition, these areas enclose curved morphological structures, whose distribution and morphology refer to salt flow. This argument is based on the correlation of the lineaments in these two areas (Figure 5b) with the seafloor morphology reported by [6,47,48]. In addition, the authors of [48] reported that, for the Central Red Sea, strongly influenced by salt flow, the morphology of the sedimentary cover on top of the moving salt often includes downslope ridges and troughs, along-slope ridges, areas of rough topography, as well as step like morphology at the flow fronts. The formation of brines in several deeps is the result of the dissolution of the Miocene evaporites upper part, and this can explain the irregular seafloor topography (vertical relief) close to the flow fronts [48]. While the origin of the downslope and along-slope ridges observed on the proposed salt flows cannot be definitely solved, their striking morphological similarity, where present, with features observed on ice glaciers or related to submarine mass movement indicates that sediment flow actually takes place near the Red Sea central trough [48]. In this work, we used the findings of [46–48] to interpret and further demonstrate how the automatically detected skeleton can assist the detection of shallow salt movements strongly related to potentially hydrocarbon-rich areas such as the wide area southwest of the Eratosthenes Seamount.

4. Discussion

4.1. Seafloor Deformation due to Halokinesis and Tectonism

The term "halokinesis" is used for autonomous salt movements due to heterogeneous density stratification in the crust [49,50]. The Mediterranean Basin is considered as the largest intracontinental deep-water basin bordered by Europe and Afro-Arabia, being one of the world's largest salt-bearing super giants (Figure 1). Two main salt geological intervals occur in the Mediterranean Basin; Triassic-Lower Jurassic salt and Upper Miocene (Messinian) salt (see Figures 8.5, 8.7, and 8.8 in [8]). Examples of seismic reflection (i.e., the most reliable ground truth method for imaging earth's crust), showing overburden and related seafloor deformation due to halokinesis, are shown in Figure 7. Figure 7a,b present migrated seismic sections from line ION-7 crossing the Ionian Basin [51], corresponding to both intrusions of Messinian and Triassic evaporites. Figure 7a images the seafloor and underlying strata in the Ionian Abyssal Plain to reveal that Messinian salt deforms the overlying Pliocene-Quaternary strata. This kind of morphology is the so-called 'cobblestone' morphology [52,53]. The term 'cobblestone' morphology is used when recognizing small-scale, topographic bathymetric features with short wavelengths, a pattern primarily associated with small scale, diapiric movements. Figure 7b presents the seafloor deformation due to diapiric movements of Triassic evaporites in the hanging-wall anticlines of pre-existing thrusts [51,54,55]. Seafloor deformation due to the diapiric movements of Triassic evaporites is clearly characterized by larger scale topographic highs compared with the deformation due to the diapiric movements of the Messinian salts. Figure 7c shows an example from Southern Crete (Ptolemy Trough), where the seafloor is only slightly deformed as a result of halokinesis [43]. In Figure 7d, very moderate seafloor deformation in the form of small-scale ridges and troughs is visible in the NE–SW trending profile located SW of the Eratosthenes Seamount [56]. At this point, we point out that the deformation of the Pliocene-Pleistocene sequence overlying the Messinian salt in salt-rich basins around the Eratosthenes Seamount in the Eastern Mediterranean has been confirmed by abundant studies over many decades [46–48,56–77].

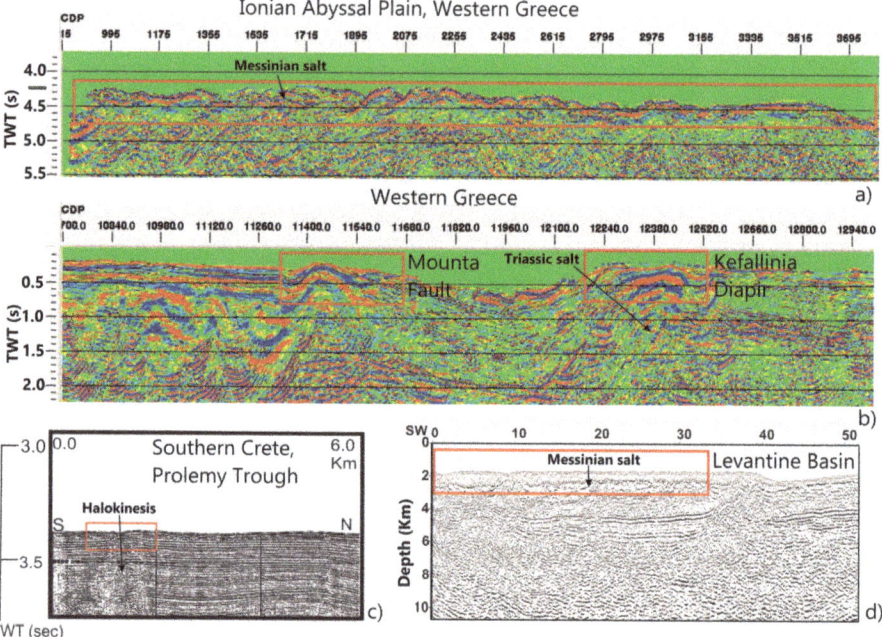

Figure 7. Seismic reflection profiles highlighting the styles of seafloor deformation due to the presence of underlying salt—see red rectangles. TWT corresponds to two way travel time. The locations of the seismic reflection sections are shown in Figure 1. These high-resolution 2D profiles image Messinian salt (**a**) in the Ionian Abyssal Plain, together with underlying Triassic evaporites; (**b**) around the Kefallinia Diapir [51]; (**c**) N–S trending high-resolution 2D profile from the Ptolemy trough, Southern Crete (Greece), imaging very moderate deformation of the seafloor owing to halokinesis [43]; (**d**) NE–SW trending high-resolution 2D profile located SW of the Eratosthenes Seamount revealing the presence of Messinian salt in the Levantine Basin [56].

The bathymetric pattern related to salt presents distinct differences when compared with the bathymetric pattern related to intense tectonic activity. In [6], a comparison of the bathymetric patterns on and north of the Eratosthenes Seamount in Eastern Mediterranean proved that the present seafloor morphology north of Eratosthenes Seamount relates to shallow salt movements, creating a chaotic pattern. On the contrary, the bathymetric pattern on the Eratosthenes Seamount relates to the action of E–W trending geological faults. Going one step forward in this work, we apply multiple filtering and skeletonization to (a) correlate the automatically detected pattern of Gavdos Rise in Southern Crete with ground truth data (Figures 5a and 6) and (b) detect patterns related to shallow salt movements southwest of the Eratosthenes Seamount (Figure 5b). Taking into account [6] and the present work, we report the following:

- Well-distinguished groups of lineaments trending to specific directions (Figures 5a and 6) characterize the bathymetric pattern related to areas mostly affected by tectonism. The detection strength of the lineaments for these groups is generally high (red and yellow colors in Figures 3a and 6);
- Morphostructures on the seafloor related to shallow salt movement (halokinesis) are recognized based on flow like characteristics such as (a) irregular seafloor topography, (b) downslope ridges and troughs, (c) presence of flow lobes (d) along-slope ridges and areas of rough topography, and (e) steplike morphology. In addition, areas with seafloor related to salt movements may be limited by distinctive, linear or curvilinear, continuous, long, and strongly detected lineaments that probably correspond to large-scale geological faults (Figure 5b, area indicated by green line).

Considerable work on the automatic extraction of geological lineaments has been published in the last years. The authors of [19] developed an algorithm that is based on tensor voting coupled Hough transform, aiming to extract geological lineaments from DEM and Landsat 8 OLI for the Loess areas in Baoji, China. Later, the authors of [78] proved that the extraction of lineaments is affected by the differences in the vertical accuracy. Specifically, the number of extracted lineaments as well as their length and density increase when the accuracy of the DEM increases. Recently, the authors of [79] managed to extract geological lineaments for a mine area in China based on wavelet edge detection and tracking by hillshade. However, all three previously mentioned works used land DEMs in the experimental phase supported by an abundance of ground truth data to validate their results. On the contrary, the present work used marine DEMs of considerably large areas in Eastern Mediterranean with limited ground truth data and not of very high resolution. However, multiple filtering and skeletonization seem to be effective for the automatic extraction of tectonic lineaments and salt-related morphostructures in bathymetric data even when bathymetric data are not of very high accuracy.

4.2. The Contribution of Multiple Filtering and Skeletonization in Marine Geology and Geophysics

Multiple filtering and skeletonization present a wide application on different problems of computer vision and pattern recognition with significant real-world applications in remote sensing [19,34,78–81], security [82], medicine [83], and computer graphics [84], especially for the identification and modelling of structures. Thus, these techniques, as previously shown, can significantly assist the interpretation of bathymetric and geophysical images to recognize seafloor and further subsurface structures. Specifically, multiple filtering and skeletonization can be applied in the following ways:

- prior to marine mapping at the stage where already available digitized data are reprocessed and evaluated, aiming to accurately specify the research targets and reduce the time and financial cost of the entire marine research;
- in digital bathymetric data acquired through remote sensing techniques or marine surveying.

The combination of multiple filtering and skeletonization techniques to identify tectonic lineaments and patterns related to evaporite movements with seafloor expression presents several advantages:

- Both are fast and simple methods with linear computational complexity concerning the image area, yielding high accuracy detections, as shown in Figures 3–6 of Section 3.
- Multiple filtering is scale and orientation invariant [78].
- The skeleton accurately reflects the shape of the original image, preserving the topological properties (homotopy) and symmetry of the object. It holds that the stability of these methods is connected to the data quality (Figures 3–5).
- Automatic identification does not require any ground truth data or training process as the popular deep learning (neural networks)-based methods [85], which cannot be applied in most of the cases, owing to unavailable or very limited ground truth data.
- The detection strength, the length, the direction of lineaments, as well as their spatial distribution are robust criteria for their sorting and further geological interpretation.

The only drawback of the proposed scheme is some possible false detections after the skeletonization process, as skeletonization is sensitive to minor boundary deformations and image noise. Some of them can be automatically removed via skeleton pruning [86], if they concern skeleton parts of well-detected medial axes. In addition, segments with unpredictable and irregular orientation can be automatically detected as outliers, if we compare their orientation with the orientation of the other detected medial axes in a neighborhood. This part can be also included in our future work.

5. Conclusions

This work demonstrates a fast and effective methodology of bathymetric analysis based on image multiple filtering and skeletonization. The experimental results with bathymetric data from the Southern Cretan offshore and the Levantine Basin (on, east, and southwest of the Eratosthenes Seamount) and comparison with ground truth data indicate the reliable, time-effective, and cost-effective performance of the proposed scheme. As a conclusion:

- The skeleton, provided by the application of the present scheme, preserves the shape and symmetry of the original lineaments. In addition, the tracing of the lineaments is accurate, taking into account their location, direction, and spatial extend. Obviously, the accuracy of lineaments automatic extraction depends on DEM's accuracy.
- Multiple filtering and skeletonization are independent in training process, so it works even in cases of limited ground truth data.
- Robust criteria, such as the strength of the detection, the length, the direction, the spatial distribution, and/or density of the lineaments, are valuable for the interpretation of the bathymetric patterns. Consequently, either seafloor morphology related with intense tectonic activity or evaporite movements with seafloor expression can be recognized and further interpreted to assist the structural analysis of potentially hydrocarbon-rich areas.

Ongoing work targets (a) the automatic erasing of false detections and the combination of this method with deep learning when more ground data are available, and (b) the prediction of possible natural gas or oil deposits in combination with related datasets.

Author Contributions: Conceptualization, E.K.; methodology, E.K. and C.P.; software, E.K. and C.P.; validation, E.K. and C.P.; formal analysis, E.K. and C.P.; investigation, E.K. and C.P.; resources, E.K. and C.P.; data curation, E.K.; writing—original draft preparation, E.K.; writing—review and editing, E.K. and C.P.; visualization, E.K. and C.P.; supervision, E.K.; project administration, E.K. All authors have read and agreed to the published version of the manuscript.

Funding: This research received no external funding.

Acknowledgments: The authors are grateful to the editor, assistant editor, and reviewers for their critical review and constructive comments. The Laboratory of Applied Geology and Hydrogeology and the Laboratory of Data Science, Multimedia, and Modelling, Hellenic Mediterranean University have supported this research.

Conflicts of Interest: The authors declare no conflict of interest.

References

1. Micallef, A.; Krastel, S.; Savini, A. (Eds.) Introduction. In *Submarine Geomorphology*; Springer: Berlin, Germany, 2018; pp. 1–9.
2. Lecours, V.; Dolan, M.F.J.; Micallef, A.; Lucieer, V.L. A review of marine geomorphometry, the quantitative study of the seafloor. *Hydrol. Earth Syst. Sci.* **2016**, *20*, 3207–3244. [CrossRef]
3. Lucieer, V.L.; Lecours, V.; Dolan, M.F.J. *Marine Geomorphometry*; MDPI: Basel, Switzerland, 2019; pp. 1–402.
4. Parsons, B.; Sclater, J.G. An analysis of the variation of ocean floor bathymetry and heat flow with age. *J. Geophys. Res.* **1977**, *82*, 803–827. [CrossRef]
5. Wessel, P.; Chandler, M.T. The spatial and temporal distribution of marine geophysical surveys. *Acta Geophys.* **2011**, *59*, 55–71. [CrossRef]
6. Kokinou, E.; Panagiotakis, C. Structural pattern recognition applied on bathymetric data from the Eratosthenes Seamount (Eastern Mediterranean, Levantine Basin). *Geo-Mar. Lett.* **2018**, *38*, 527–540. [CrossRef]
7. Camerlenghi, A. Drivers of Seafloor Geomorphic Change. In *Submarine Geomorphology*; Micallef, A., Krastel, S., Savini, A., Eds.; Springer: Cham, Switzerland, 2018; pp. 135–159.
8. Belenitskaya, G. *Salt Systems of the Earth: Distribution, Tectonic and Kinematic History, Salt-Naphthids Interrelations, Discharge Foci, Recycling*; Wiley Online Library: Hoboken, NJ, USA, 2018; pp. 1–693.
9. Kweon, I.S.; Kanade, T. Extracting topographic terrain features from elevation maps. *CVGIP Image Underst.* **1994**, *59*, 171–182. [CrossRef]
10. Miliaresis, G.C.; Argialas, D. Segmentation of physiographic features from the global digital elevation model/gtopo30. *Comput. Geosci.* **1999**, *25*, 715–728. [CrossRef]
11. Dinesh, S. Extraction of mountains from digital elevation models using mathematical morphology. *GIS Dev. Malays.* **2006**, *1*, 16–19.

12. Iwahashi, J.; Pike, R.J. Automated classifications of topography from dems by an unsupervised nested-means algorithm and a three-part geometric signature. *Geomorphology* **2007**, *86*, 409–440. [CrossRef]
13. Booth, A.M.; Roering, J.J.; Perron, J.T. Automated landslide mapping using spectral analysis and high-resolution topographic data: Puget sound Lowlands, Washington, and Portland hills, Oregon. *Geomorphology* **2009**, *109*, 132–147. [CrossRef]
14. Obu, J.; Podobnikar, T. Algorithm for karst depression recognition using digital terrain models. *Geod. Vestn.* **2013**, *57*, 260–270. [CrossRef]
15. Euillades, L.D.; Grosse, P.; Euillades, P.A. Netvolc: An algorithm for automatic delimitation of volcano edifice boundaries using dems. *Comput. Geosci.* **2013**, *56*, 151–160. [CrossRef]
16. Panagiotakis, C.; Kokinou, E. Linear Pattern Detection of Geological Faults via a Topology and Shape Optimization Method. *IEEE J. Sel. Top. Appl. Earth Obs. Remote Sens.* **2015**, *8*, 3–11. [CrossRef]
17. Micheal, A.A.; Vani, K. Automatic mountain detection in lunar images using texture of dtm data. *Comput. Geosci.* **2015**, *82*, 130–138. [CrossRef]
18. Panagiotakis, C.; Kokinou, E. Unsupervised Detection of Topographic Highs with Arbitrary Basal Shapes Based on Volume Evolution of Isocontours. *Comput. Geosci.* **2017**, *102*, 22–33. [CrossRef]
19. Han, L.; Liu, Z.; Ning, Y.; Zhao, Z. Extraction and analysis of geological lineaments combining a DEM and remote sensing images from the northern Daoji loess area. *Adv. Space Res.* **2018**, *62*, 2480–2493. [CrossRef]
20. Florinsky, I.V. Quantitative topographic method of fault morphology recognition. *Geomorphology* **1996**, *16*, 103–119. [CrossRef]
21. Florinsky, I.V. *Digital Terrain Analysis in Soil Science and Geology*, 2nd ed.; Elsevier: Amsterdam, The Netherlands, 2016; pp. 1–486.
22. Warren, J.K. *Evaporites, a Geological Compendium*, 2nd ed.; Springer: Berlin, Germany, 2016; pp. 1–1813.
23. Hudec, M.R.; Jackson, M.P.A. Terra infirma: Understanding salt tectonics. *Earth-Sci. Rev.* **2007**, *82*, 1–28. [CrossRef]
24. Hobbs, W.H. Lineaments of Atlantic Border region. *Bull. Geol. Soc. Am.* **1904**, *15*, 483–506. [CrossRef]
25. O'Leary, D.W.; Friedman, J.D.; Pohn, H.A. Lineament, linear, lineation: Some proposed new standards for old terms. *Bull. Geol. Soc. Am.* **1976**, *87*, 1463–1469. [CrossRef]
26. Makarov, V.I. Lineaments: Problems and trends of studies by remote sensing techniques. *Izv. Vuz. Geol. Razv.* **1981**, *4*, 109–115. (In Russian)
27. Florinsky, I.V. Global Lineaments: Application of Digital Terrain Modelling. In *Advances in Digital Terrain Analysis. Lecture Notes in Geoinformation and Cartography*; Zhou, Q., Lees, B., Tang, G., Eds.; Springer: Berlin/Heidelberg, Germany, 2008; pp. 365–382.
28. Berthou, P. EMODNET-the European marine observation and data network. *Eur. Sci. Found. Mar. Board* **2008**, *10*.
29. Benkhelil, J.; Mart, Y.; Mascle, J.; Woodside, J.M.; The Prismed II Scientific Party. Recent morphostructural evolution of the Eratosthenes Seamount: Initial Results of the Prismed II Cruise. European Union of Geosciences/EUG 10, Strasbourg (France). *J. Conf. Abs.* **1999**, *4*, 756, Terra Abstracts 11.
30. CIESM/Ifremer Medimap Group; Loubrieu, B.; Mascle., J. Morpho-bathymetry of the Mediterranean Sea. CIESM: Monte Carlo, Monaco, 2008.
31. Huebscher, C. *RV MARIA S. MERIAN, Cruise Report MSM14/L3 Eratosthenes Seamount/Eastern Mediterranean Sea 2010*; DFG Senatskommission für Ozeanographie: Bremen, Germany, 2012.
32. U.S. Department of Commerce, National Oceanic and Atmospheric Administration, National Geophysical Data Center. 2-Minute Gridded Global Relief Data (ETOPO2v2). 2006. Available online: http://www.ngdc.noaa.gov/mgg/fliers/06mgg01.html (accessed on 12 May 2020).
33. Kokinou, E.; Panagiotakis, C.; Sarris, A. Application of Skeletonization on Geophysical Images. In Proceedings of the XIX Congress of the Carpathian Balkan Geological Association, Thessaloniki, Grecce, 23–26 September 2010; Volume 99, pp. 338–343.
34. Panagiotakis, C.; Kokinou, E.; Sarris, A. Curvilinear Structure Enhancement and Detection in Geophysical Images. *IEEE Trans. Geosci. Remote Sens.* **2011**, *49*, 2040–2048. [CrossRef]
35. Alves, T.M.; Kokinou, E.; Zodiatis, G.; Lardner, R.; Panagiotakis, C.; Radhakrishnan, H. Modelling of oil spills in confined maritime basins: The case for early response in the Eastern Mediterranean Sea. *Environ. Pollut.* **2015**, *206*, 390–399. [CrossRef] [PubMed]

36. Alves, T.; Kokinou, E.; Zodiatis, G.; Lardner, R. Hindcast, GIS and susceptibility modelling to assist oil spill clean-up and mitigation on the southern coast of Cyprus (Eastern Mediterranean). *Deep-Sea Res. Part II* **2016**, *133*, 159–175. [CrossRef]
37. Alves, T.M.; Kokinou, E.; Zodiatis, G.; Radhakrishnan, H.; Panagiotakis, C.; Lardner, R. Multidisciplinary oil spill modeling to protect coastal communities and the environment of the Eastern Mediterranean Sea. *Sci. Rep.* **2016**, *6*, 36882. [CrossRef]
38. Kokinou, E. Geomorphologic features of the marine environment in Eastern Mediterranean using a modern processing approach. In Proceedings of the IAMG 2015, the 17th Annual Conference of the International Association for Mathematical Geosciences, Freiberg, Germany, 5–13 September 2015; Schaeben, H., Tolosana Delgado, R., van den Boogaart, G.K., van den Boogaart, R., Eds.; IAMG: Freiberg, Germany, 2015; pp. 436–445.
39. Andronikidis, N.; Kokinou, E.; Vafidis, A.; Kamberis, E.; Manoutsoglou, E. Deformation patterns in the southwestern part of the Mediterranean Ridge (South Matapan Trench, Western Greece). *Mar. Geophys. Res.* **2018**, *39*, 475–490. [CrossRef]
40. Toulia, E.; Kokinou, E.; Panagiotakis, C. The contribution of pattern recognition techniques in geomorphology and geology: The case study of Tinos Island (Cyclades, Aegean, Greece). *Eur. J. Remote Sens.* **2018**, *51*, 88–99. [CrossRef]
41. Canny, J. A computational approach to edge detection. *IEEE Trans. Pattern Anal. Mach. Intell.* **1986**, *8*, 679–698. [CrossRef]
42. Kermad, C.D.; Chehdi, K. Automatic image segmentation system through iterative edge-region co-operation. *Image Vis. Comput.* **2002**, *8*, 541–555. [CrossRef]
43. Kokinou, E.; Alves, T.M.; Kamberis, E. Structural decoupling on a convergent forearc setting (Southern Crete, Eastern Mediterranean). *Geol. Soc. Am. Bull.* **2012**, *124*, 1352–1364. [CrossRef]
44. IGME (Institute of Mineral and Geological Exploration). *Seismotectonic Map of Greece. Scale 1:500000*; IGME: Athens, Greece, 1989.
45. Manta, K.; Rousakis, G.; Anastasakis, G.; Lykousis, V.; Sakellariou, D.; Panagiotopoulos, I.P. Sediment transport mechanisms from the slopes and canyons to the deep basins south of Crete Island (southeast Mediterranean). *Geo-Mar. Lett.* **2019**, *39*, 295–312. [CrossRef]
46. Mascle, J.; Sardou, O.; Loncke, L.; Migeon, S.; Camera, L.; Gaullier, V. Morphostructure of the Egyptian continental margin: Insights from swath bathymetry surveys. *Mar. Geophys. Res.* **2006**, *27*, 49–59. [CrossRef]
47. Schattner, U. What triggered the early-to-mid Pleistocene tectonic transition across the entire eastern Mediterranean? *Earth Planet. Sci. Lett.* **2010**, *289*, 539–548. [CrossRef]
48. Feldens, P.; Mitchell, N. Salt flows in the Central Red Sea. In *The Red Sea: The Formation, Morphology, Oceanography and Environment of a Young Ocean Basin*; Rasul, N., Stewart, I., Eds.; Springer: Berlin/Heidelberg, Germany, 2015; pp. 205–218.
49. Trusheim, F. Über Halokinese und ihre Bedeutung für die strukturelle Entwicklung Norddcutschlands. *Z. Deutch. Geol. Ges.* **1957**, *198*, 111–151.
50. Trusheim, F. Mechanism of salt migration in northern Germany. *AAPG Bull.* **1960**, *44*, 1519–1540.
51. Kokinou, E.; Kamberis, E.; Vafidis, A.; Monopolis, D.; Ananiadis, G.; Zelilidis, A. Deep seismic reflection data from offshore Western Greece: A new crustal model for the Ionian Sea. *J. Pet. Geol.* **2005**, *28*, 81–98. [CrossRef]
52. Hersey, J.B. Sedimentary basins of the Mediterranean Sea. In *Submarine Geology and Geophysics, Proceedings of the 17th Symposium of the Colston Research Society, London, UK, 5–9 April 1965*; Whitard, W.F., Bradshaw, W., Eds.; Butterworths: London, UK, 1965; pp. 75–91.
53. Hsu, K.J.; Cita, M.B. The origin of the Mediterranean Evaporite. In *Initial Reports of the Deep Sea Drilling Project*; U.S. Government Printing Office: Washington, DC, USA, 1973; Volume 13, pp. 1203–1231.
54. Monopolis, D.; Bruneton, A. Ionian Sea (Western Greece): Its structural outline deduced from drilling and geophysical data. *Tectonophysics* **1982**, *83*, 227–242. [CrossRef]
55. Brooks, M.; Ferentinos, G. Tectonics and Sedimentation in the Gulf of Corinth and Zakynthos and Kefallinia channels, western Greece. *Tectonophysics* **1984**, *101*, 25–54. [CrossRef]
56. Kopf, A.; Vidal, N.; Klaeschen, D.; von Huene, R.; Krasheninnikov, V.A. Multi-channel seismic profiles across Eratosthenes Seamount and the Florence Rise reflecting the incipient collision between Africa and Eurasia near the island of Cyprus, Eastern Mediterranean. In *Geological Framework of the Levant, Volume II: The Levantine Basin and Israel*; Hall, J.K., Krasheninnikov, V.A., Hirsch, F., Benjamini, C., Flexer, A., Eds.; Historical Productions-Hall: Jerusalem, Israel, 2005; pp. 57–72.

57. Neev, D.; Almagor, G.; Arad, V.; Ginzburg, A.; Hall, J.K. The geology of the southeastern Mediterranean Sea. *Geol. Surv. Isr. Bull.* **1976**, *68*, 1051.
58. Ben-Avraham, Z. The structure and tectonic setting of the Levant continental margin, Eastern Mediterranean. *Tectonophysics* **1978**, *46*, 313–331. [CrossRef]
59. Garfunkel, Z.; Arad, V.; Almagor, G. The Palmahim disturbance and its regional setting. *Geol. Surv. Isr. Bull.* **1979**, *72*, 56.
60. Mart, Y.; Ben-Gai, Y. Some depositional patterns at continental margin of southeaster Mediterranean Sea. *AAPG Bull.* **1982**, *66*, 460–470.
61. Almagor, G.; Hall, J.K. Morphology of the continental margin off north-central Israel. *Isr. J. Earth Sci.* **1983**, *32*, 75–82.
62. Garfunkel, Z. Large-scale submarine rotational slumps and growth faults in the eastern Mediterranean. *Mar. Geol.* **1984**, *55*, 304–324. [CrossRef]
63. Frey-Martinez, J.; Cartwright, J.A.; Hall, B. 3D seismic interpretation of slump complexes: Examples from the continental margin of Israel. *Basin Res.* **2005**, *17*, 83–108. [CrossRef]
64. Gradmann, S.; Huebscher, C.; Ben-Avraham, Z.; Gajewski, D.; Netzband, G. Salt tectonics off northern Israel. *Mar. Pet. Geol.* **2005**, *22*, 597–611. [CrossRef]
65. Bertoni, C.; Cartwright, J.A. Controls on the basinwide architecture of late Miocene (Messinian) evaporites on the Levant margin (Eastern Mediterranean). *Sediment. Geol.* **2006**, *188*, 93–114. [CrossRef]
66. Bertoni, C.; Cartwright, J.A. Major erosion at the end of the Messinian Salinity Crisis: Evidence from the Levant Basin, Eastern Mediterranean. *Basin Res.* **2007**, *19*, 1–18. [CrossRef]
67. Loncke, L.; Gaullier, V.; Mascle, J.; Vendeville, B.; Camera, L. The Nile deep-sea fan: An example of interacting sedimentation, salt tectonics, and inherited subsalt paleotopographic features. *Mar. Pet. Geol.* **2006**, *23*, 297–315. [CrossRef]
68. Netzeband, G.; Huebscher, C.; Gajewski, D. The structural evolution of the Messinian evaporites in the Levantine Basin. *Mar. Geol.* **2006**, *230*, 249–273. [CrossRef]
69. Mart, Y.; Ryan, W. The Levant slumps and the Phoenician structures: Collapse features along the continental margin of the southeastern Mediterranean Sea. *Mar. Geophys. Res.* **2007**, *28*, 297–307. [CrossRef]
70. Cartwright, J.A.; Jackson, M.P.A. Initiation of gravitational collapse of an evaporate basin margin: The Messinian saline giant, Levant Basin, eastern Mediterranean. *Geol. Soc. Am. Bull.* **2008**, *120*, 399–413. [CrossRef]
71. Clark, I.R.; Cartwright, J.A. Interactions between submarine channel systems and deformation in deepwater fold belts: Examples from the Levant Basin, Eastern Mediterranean Sea. *Mar. Pet. Geol.* **2009**, *26*, 1465–1482. [CrossRef]
72. Gvirtzman, Z.; Reshef, M.; Buch-Leviatan, O.; Ben-Avraham, Z. Intense salt deformation in the Levant Basin in the middle of the Messinian Salinity Crisis. *Earth Planet. Sci. Lett.* **2013**, *379*, 108–119. [CrossRef]
73. Reiche, S.; Huebscher, C. The Hecataeus Rise, easternmost Mediterranean: A structural record of Miocene-Quaternary convergence and incipient continent-continent-collision at the African-Anatolian plate boundary. *Mar. Pet. Geol.* **2015**, *67*, 368–388. [CrossRef]
74. Symeou, V.; Homberg, C.; Nader, F.H.; Darnault, R.; Lecomte, J.C.; Papadimitriou, N. Longitudinal and Temporal Evolution of the Tectonic Style Along the Cyprus Arc System, Assessed Through 2-D Reflection Seismic Interpretation. *Tectonics* **2018**, *37*, 30–47. [CrossRef]
75. Rouchy, J.M.; Caruso, A. The Messinian salinity crisis in the Mediterranean basin: A reassessment of the data and an integrated scenario. *Sediment. Geol.* **2006**, *188*, 35–67. [CrossRef]
76. Manzi, V.; Gennari, R.; Lugli, S.; Roveri, M.; Scafetta, N.; Schreiber, B.C. High frequency cyclicity in the Mediterranean Messinian evaporites: Evidence for solar–lunar climate forcing. *J. Sediment. Res.* **2012**, *82*, 991–1005. [CrossRef]
77. Roveri, M.; Flecker, R.; Krijgsman, W.; Lofi, J.; Lugli, S.; Manzi, V.; Sierro, F.J.; Bertini, A.; Camerlenghi, A.; De Lange, G.; et al. The Messinian Salinity Crisis: Past and future of a great challenge for marine sciences. *Mar. Geol.* **2014**, *352*, 25–58. [CrossRef]
78. Soliman, A.; Han, L. Effects of vertical accuracy of digital elevation model (DEM) data on automatic lineaments extraction from shaded DEM. *Adv. Space Res.* **2019**, *64*, 603–622. [CrossRef]
79. Xu, J.; Wen, X.; Zhang, H.; Luo, D.; Li, J.; Xu, L.; Yu, M. Automatic extraction of lineaments based on wavelet edge detection and aided tracking by hillshade. *Adv. Space Res.* **2020**, *65*, 506–517. [CrossRef]

80. Gournia, C.; Fakiris, E.; Geraga, M.; Williams, D.P.; Papatheodorou, G. Automatic Detection of Trawl-Marks in Sidescan Sonar Images through Spatial Domain Filtering, Employing Haar-Like Features and Morphological Operations. *Geosciences* **2019**, *9*, 214. [CrossRef]
81. Parker, J.R. *Algorithms for Image Processing and Computer Vision*, 2nd ed.; Wiley Publishing: Indianapolis, IN, USA, 2010; ISBN 0470643854.
82. Bounneche, M.D.; Boubchir, L.; Bouridane, A.; Nekhoul, B.; Ali-Chérif, A. Multi-spectral palmprint recognition based on oriented multiscale log-Gabor filters. *Neurocomputing* **2016**, *205*, 274–286. [CrossRef]
83. Panagiotakis, C.; Argyros, A. Region-based Fitting of Overlapping Ellipses and its Application to Cells Segmentation. *Image Vis. Comput.* **2020**, *93*, 103810. [CrossRef]
84. Panagiotakis, C.; Tziritas, G. Snake terrestrial locomotion synthesis in 3D virtual environments. *Vis. Comput.* **2006**, *22*, 86–98. [CrossRef]
85. Zhu, X.X.; Tuia, D.; Mou, L.; Xia, G.S.; Zhang, L.; Xu, F.; Fraundorfer, F. Deep learning in remote sensing: A comprehensive review and list of resources. *IEEE Geosci. Remote Sens. Mag.* **2017**, *5*, 8–36. [CrossRef]
86. Liu, H.; Wu, Z.H.; Zhang, X.; Hsu, D.F. A skeleton pruning algorithm based on information fusion. *Pattern Recognit. Lett.* **2013**, *34*, 1138–1145. [CrossRef]

© 2020 by the authors. Licensee MDPI, Basel, Switzerland. This article is an open access article distributed under the terms and conditions of the Creative Commons Attribution (CC BY) license (http://creativecommons.org/licenses/by/4.0/).

Article

Integrating Remote Sensing and Street View Images to Quantify Urban Forest Ecosystem Services

Elena Barbierato [1], Iacopo Bernetti [1,*], Irene Capecchi [1] and Claudio Saragosa [2]

[1] Department of Agriculture, Food, Environment and Forestry University of Florence, Piazzale delle Cascine 18, 50144 Firenze, Italy; elena.barbierato@unifi.it (E.B.); irene.capecchi@unifi.it (I.C.)
[2] Department of Architecture, University of Florence, Via della Mattonaia, 14, 50121 Firenze, Italy; claudio.saragosa@unifi.it
* Correspondence: iacopo.bernetti@unifi.it; Tel.: +39-3298603416

Received: 7 December 2019; Accepted: 14 January 2020; Published: 19 January 2020

Abstract: There is an urgent need for holistic tools to assess the health impacts of climate change mitigation and adaptation policies relating to increasing public green spaces. Urban vegetation provides numerous ecosystem services on a local scale and is therefore a potential adaptation strategy that can be used in an era of global warming to offset the increasing impacts of human activity on urban environments. In this study, we propose a set of urban green ecological metrics that can be used to evaluate urban green ecosystem services. The metrics were derived from two complementary surveys: a traditional remote sensing survey of multispectral images and Laser Imaging Detection and Ranging (LiDAR) data, and a survey using proximate sensing through images made available by the Google Street View database. In accordance with previous studies, two classes of metrics were calculated: greenery at lower and higher elevations than building facades. In the last phase of the work, the metrics were applied to city blocks, and a spatially constrained clustering methodology was employed. Homogeneous areas were identified in relation to the urban greenery characteristics. The proposed methodology represents the development of a geographic information system that can be used by public administrators and urban green designers to create and maintain urban public forests.

Keywords: urban forest; landscape metrics; LiDAR; aerial images; street view images; semantic segmentation; convolutional neural network (CNN); spatial clustering

1. Introduction

Holistic tools to assess the health impacts from climate change mitigation/adaptation policies enacted to increase the amount of public green spaces are urgently needed. Urban vegetation provides numerous ecosystem services on a local scale and is therefore a potential adaptation strategy that can be monopolized in the era of global warming, and it can also offset the increasing impacts of human activity on the urban environment.

In this respect, on their review on health and climate relating to ecosystem services provided by street trees in an urban environment, Salmon et al. [1] stated that, "Our review, in agreement with other papers in the ecosystem services (ESS) literature...has also highlighted the importance of scale when determining the effect of trees on climate and health. Whilst much of the research to date has focused on the regional and urban scale effects of vegetation on climate and health, it is much less clear what the impacts of street trees are at local scales where the result of the intervention is most clearly felt".

On a local scale, the presence of street trees can modify indoor temperatures by shading buildings and can significantly reduce the risk of indoor overheating [2]. An empirical study [3] conducted on a project scale using direct measurements obtained with an infrared camera showed that the degree of foliage shading from rows of trees on building facades may decrease surface temperatures by up to 9 °C.

Streiling and Matzarakis [4] found that clustering trees into lines or small groups interspersed with open areas can help reduce the radiative load, provide shade, and allow long-wave cooling at night. Growing large and broad trees with dense canopies can be considered in streets that have a low height/width ratio, while taller narrower trees can be grown in streets that have a high height/width ratio [5]. On a local scale, the characteristics of tree canopy, tree density, and proximity to other urban structures influence the ability of plants to remove pollutants [6,7] Nilsson et al. [8] showed that in the ecosystem service of traffic noise attenuation, the urban green characteristics on a local scale (depth, width, and stem diameter of a tree belt) play a fundamental role. In addition, with respect to cultural ecosystem services, a laboratory psychometric study was conducted in which three-dimensional (3D) videos produced in a laboratory contained urban green coverage at eye level ranging from 0% to 70%: videos that contained a higher level of public greenery elicited a greater self-reported stress reduction [9].

To transpose the results of previous research to other cities, specific ecological metrics on a local scale are needed, and a combination of traditional remote sensing and so-called "proximate sensing" appears to be a good candidate in this respect. The traditional field of remote sensing uses overhead images of distant scenes to derive geographic information, and proximate sensing uses ground-level images of objects and scenes that are nearby [10].

When classifying land cover, urban tree cover has traditionally been quantified using long-range, remotely-sensed image processing, such as satellite imagery (LANDSAT), ortho-aerial photographs or, more recently, Laser Imaging Detection and Ranging (LiDAR) [11,12], or by employing data derived from field surveys [13]. Therefore, although the ecological metrics calculated from remote images can be useful to quantify some ecosystem services, they are not applicable for assessing ecosystem services that depend on street-level urban greening measures. Earth observation data, such as multispectral satellite images, have long been used to classify green open spaces in cities. Despite the large increase in spatial resolution, urban green classification from aerial images is still considered a difficult task, since only the upper part of plants can be captured from the nadir's point of view. For this reason, high resolution satellite images are useful for classifying urban green with a wide spatial extension, such as urban forests, green parks and gardens, but they are not efficient for urban green scattered or in rows. In fact, the lack of detail at ground level makes it difficult to detect the characteristics deriving from the shape of the foliage of the plants. Therefore, the urban green classification at street level is still based on a labor-intensive territorial survey, which is inefficient and expensive. Fortunately, the growing accessibility to different geo-tagged data sources allows to merge remote sensing images with data of different modalities and observations. For example, Google Street View (GSV) offers users street-level panoramic images captured in thousands of cities around the world, which allows you to observe street scenes in big cities and thus provides instant detection capabilities and detail at the level of the soil that lack in aerial imagery. Recently, Li et al. [14,15] developed a novel Google Street View (GSV)-based method to study the distribution of street greenery. Unlike green metrics derived from remotely-sensed data, the GSV-based method quantifies street greenery using a perspective from the ground, which better represents the distribution of street greenery projected on building facades or on street pavements. The aims of previous work have been to analyze the relationship between perceived safety and green vegetation characteristics [14,16], quantify the sky view factor (SVF) from street-level imagery, and to assess environmental inequalities in terms of different types of urban greenery [17].

The aim of this current study is to create a general-purpose set of ecological metrics by combining remote sensing and proximate sensing (Street View) approaches with data retrieved from GSV, to quantify urban forest ecosystem services and provide a widely transferable methodology. In this respect, remote sensing metrics were calculated by combining high-resolution multispectral images and LiDAR data to produce indices at different altitudes with respect to the ground. The ecological metrics from proximate sensing were then calculated by semantic segmentation using pretrained deep neural networks. To estimate the validity of this approach, a set of ecological metrics was used to classify contiguous homogeneous areas of a city through a spatial clustering algorithm [18,19]. Figure 1 presents a diagram of the workflow.

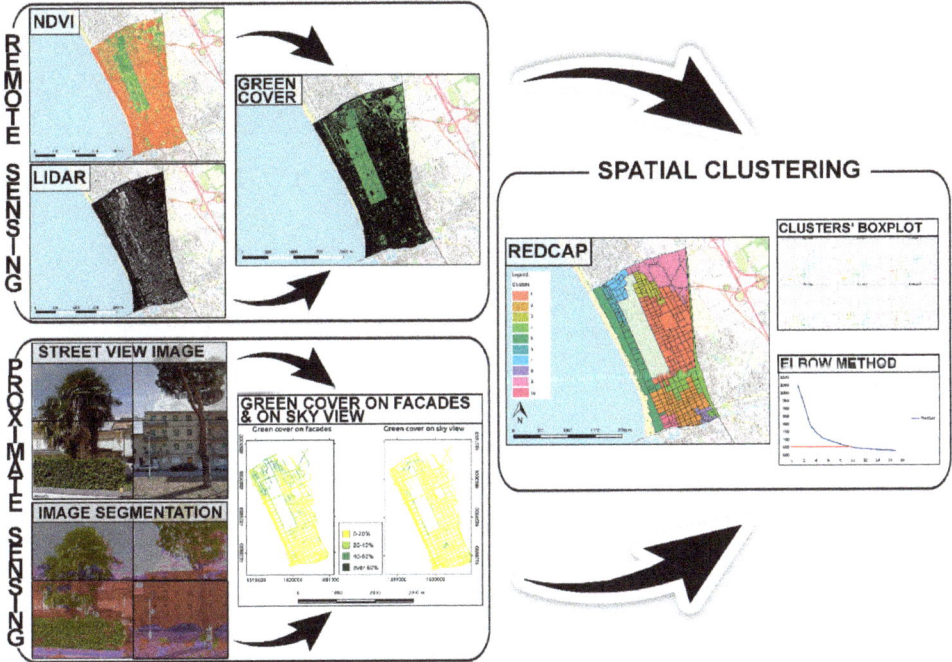

Figure 1. Workflow diagram.

2. Materials and Methods

2.1. Study Area and Data Sources

2.1.1. Study Area

Our research was conducted in the city of Viareggio in northern Tuscany, Italy. The study area has boundary coordinates (datum WGS84, projection Universal Transverse of Mercator (UTM), zone 32) Nord min = 598,681, Nord max = 601,019, East min = 4,857,573, East max = 4,860,971, and mean latitude = 43.87911° N.

Viareggio has a population of over 62,000 and is a seaside tourist town on the coast of the Ligurian Sea. It is characterized by orthogonal streets that form rectangular blocks, and the building types are terraced houses (with one or two floors), villas, and hotels. Urban greenery is widespread throughout the city, but it has different typologies. In the northern area, there is a greater percentage of greenery (hedges and rows of large trees higher than the facades of buildings), and in the southern area, small trees are spaced further apart and are at a lower height than the facades of buildings (Figure 2). The perimeter of the study area is defined by both natural and artificial borders: in the north, east, south, and west are the Fosso dell'Abate waterway, the railway, the Burlamacco canal, and the Ligurian Sea, respectively.

The study area covers an area of 3,555,104 square meters with 768,434 square meters of urban green and the remaining part by artificial surfaces.

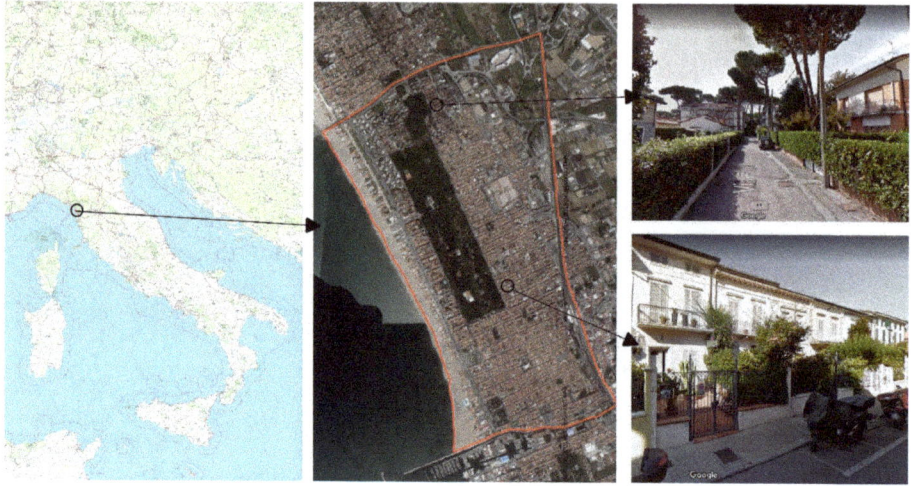

Figure 2. Study Area.

2.1.2. Data Sources

The input data were both cartographic information and remote sensing data. Remote sensing data were downloaded from a database of the Tuscan region, and vegetation cover data were obtained from photogrammetric processing of seven aerial multispectral frames (four bands: red, green and blue + near infrared regions (NIR) acquired on October 2013 using an UltraCam Xp (Vexcel) digital metric camera (resolution of 0.2 × 0.2 m). UltraCam Xp simultaneously collects light from five different spectral bands. The spectral sensitivity of red, green, blue, and near infrared and the panchromatic channel from 410 nm to 690 nm, RED from 580 nm to 700 nm, GREEN from 480 nm to 630 nm, BLUE from 410 nm to 570 nm, and NIR from 690 nm to 1000 nm.

The heights of vegetation and buildings were derived from 9 LiDAR image on 2006 (resolution of 1 × 1 m). The LiDAR data were provided by the Italian Ministry of the Environment, Land, and Sea. The points acquired from this survey have an altimetric accuracy of ±15 cm and a planimetric accuracy is ±30 cm. In this work, the data available by the geographical portal of the Tuscan region with a resolution of 1 × 1 m were used. That resolution was considered satisfactory for the objectives of the work. If necessary, however, the proposed method could therefore also be applied to more detailed scales.

The other cartographic data were derived from a topographic regional database in 2013 with detail on a scale of 1:2000. All the input raster (multispectral images and city blocks raster) were aligned at a 1 × 1 m resolution using the nearest neighbor algorithm.

2.1.3. The City Block

In this work, the reference cartographic element is the city block, as it is the central element of urban planning and urban design, and the basic building block of an urban city. We analyzed vegetation within city blocks. According to the Oxford Dictionary definition, a city block is the smallest area that is surrounded by four streets, usually containing several buildings and vegetation, and urban block boundaries are frequently used to define units for extracting metrics from remotely-sensed data [20]. In according to the definition of Oxford Dictionary, a city block is the smallest area that is surrounded by four streets, usually containing several buildings and vegetation. The blocks were obtained by clipping the study area using OpenStreetMap roads.

2.2. Proximate Sensing Landscape Metrics

2.2.1. Google Street View (GSV) Images Collection

Leung and Newsam use the term "Proximate Sensing" to describe a more comprehensive framework that uses ground level images of nearby objects and scenes to automatically map what-is-where on the surface of the Earth, similar to how remote sensing uses overhead images [21]. In this study, we used GSV images downloaded using the GSV static imageApplication programming interface API.

By specifying different parameters in the API, users can download GSV images with different fields of view, heading angles, and pitch angles. In this respect, heading indicates the compass heading of the camera, (heading values range from 0 to 360), pitch specifies the up or down angle of the camera relative to the street view vehicle, and the field of view determines the horizontal field view of the image. These parameters were used to define the ecological metrics of the streetscape.

According to literature, the ecological services of urban green areas are linked to two effects: the shading of foliage on the facades of buildings and the coverage of the sky view of the street [1,3–5,7]. For this reason, we defined two ecological metrics as follows: the percentage of green cover on facades (GCF), which is defined as Equation (1):

$$GCF = \left(\frac{\sum \text{number of pixels classified a street inside } FOV_{facades}}{\sum \text{total number of pixels inside } FOV_{facades}} \right), \quad (1)$$

And percentage of green cover on sky view of the street (GCS), which is defined as Equation (2):

$$GCS = \left(\frac{\sum \text{number of pixels classified a street inside } FOV_{sky}}{\sum \text{total number of pixels inside } FOV_{sky}} \right). \quad (2)$$

where $FOV_{facades}$ and FOV_{sky} are the field of view of the image of the facades of buildings and the field of view of the sky view of the street respectively, and l and r are the left and right sides of the street.

Images from all image collection points along city roads were downloaded every 15 m along each roadway, although there were no data for some GSV sampling points, road segments, and areas of the city for various reasons (such as corrupt or missing data or areas with no-coverage). Notwithstanding those instances, the sampling regime covered the full extent of the cities' official boundaries. Four ecological metrics can be extracted for each GSV sampling point: two for the right side and two for the left side.

The following methodology was used to define the parameters necessary for extracting the four GSV images: first, we linked the distances of headings and heights relative to the buildings on the right and left with the GSV sampling points. The buildings' heights were calculated using a map overlay operation between LiDAR geodata and the building layer of the Open Street Map geodatabase. The procedure used to link the geometric parameters of the streetscape (headings, heights, and distances) with the GSV sampling points (Figure 3a) was conducted using Geographic Resources Analysis Support System GRASS software and is available as supplementary material (file grassProcedure.txt). The $FOV_{facades}$ and the FOV_{sky} were then calculated using the following formula (Figure 3b), respectively:

$$FOV_{facade}^{l,r} = 2 \cdot \tan^{-1}\left(\frac{h_{facades}^{l,r} - h_{google}}{l_{street}^{l,r}} \right) \quad (3)$$
$$FOV_{sky}^{l,r} = 90 - FOV_{facade}^{l,r}$$

Finally, the pitch angles relative to the two metrics were obtained using the following:

$$pitch_{facade}^{l,r} = 0$$
$$pitch_{sky}^{l,r} = \frac{FOV_{facade}^{l,r}}{2} + \frac{FOV_{sky}^{l,r}}{2} \cdot \quad (4)$$

Each digital photograph (red-green-blue color channel jpeg image) was acquired from the GSV API at a resolution of 400 by 400 pixels. The images were downloaded via the "Googleway" library using the statistical software, R. The procedure used is available as supplementary material (file R_procedure.R).

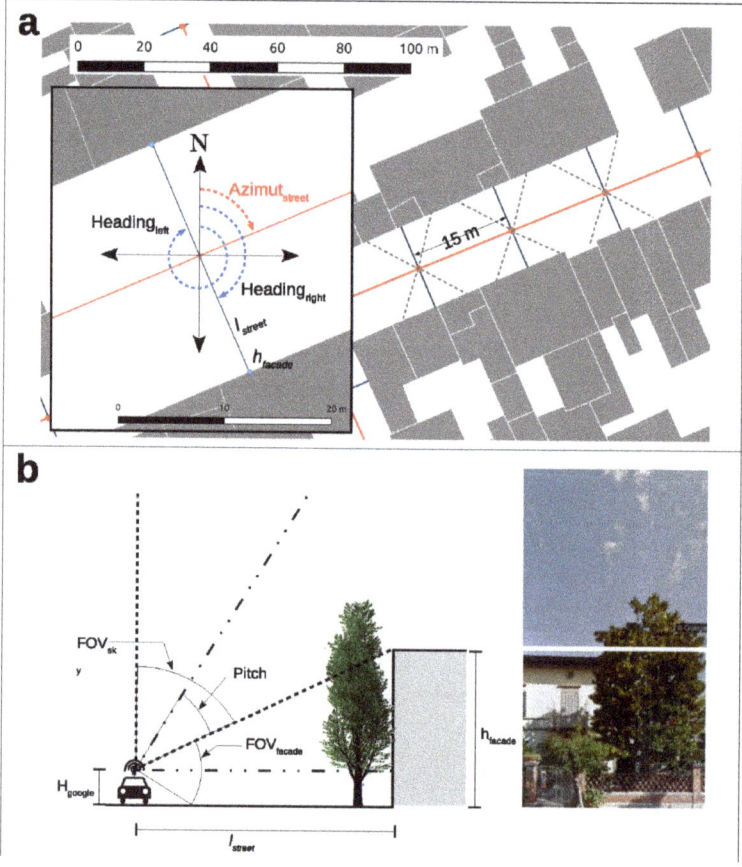

Figure 3. Sampling procedure: (**a**) sampling points, (**b**) Pitch and Field of View (FOV).

2.2.2. Image Segmentation

We estimated the total area covered by trees in each image by applying semantic segmentation using deep learning [22]. A semantic segmentation network classifies every pixel in an image, which results in an image segmented by class. In this phase of the work, we used the pre-trained network of Matrix Laboratory MATLAB software based on the Deeplabv3+ network, which is one type of convolutional neural network (CNN) designed for semantic image segmentation [23], with weights initialized by a pre-trained ResNet-18 network. ResNet-18 is an efficient network that is well suited to applications that have limited computing resources. The network was trained using the CamVid dataset [24] from the University of Cambridge, which is a collection of images containing street-level views obtained while driving, and it provides pixel-level labels for 32 semantic classes including car, pedestrian, and road. To make the training easier, the 32 original classes in CamVid were grouped into 11 classes as follows: "Sky", "Building", "Pole", "Road", "Pavement", "Tree", "SignSymbol", "Fence", "Car", "Pedestrian", and "Cyclist".

To validate the network, we extracted 200 random images from those downloaded with Street View API, manually segmented them, and then used them as a validation set to evaluate the performances of the pretrained network. Deeplab v3+ is trained using 60% of the images from the dataset. The rest of the images are split evenly in 20% and 20% for validation and testing, respectively.

2.3. Remote Sensing Landscape Metrics

The remote sensing data were used to obtain the coverage and height of vegetation. The urban vegetation coverage was identified through an analysis of the normalized difference vegetation index (NDVI). As reported in the literature [25,26], since only healthy vegetation was included in the study, it was extracted with respect to the NDVI having a value greater than or equal to 0.2. The result was presented as a Boolean map with a resolution of 1 m (which is similar to LiDAR data), in which the value of 0 indicates an absence of vegetation, while a value of 1 indicates the presence of vegetation.

Since the urban green area can be characterized by various types of vegetation with differing heights, shapes, and ecological functions, we distinguished two types according to the average height value of the buildings in each city block. To obtain the height of the vegetation, we overlaid the NDVI binary and normalized digital surface model (nDSM) generated from LIDAR data, which provided a raster map divided into two height classes (Figure 4): the first class is green cover on facades, and is represented by a value less than (or equal to) the average height of the buildings, and the second is green cover seen on a sky view, and is represented by the average value of the height of the buildings.

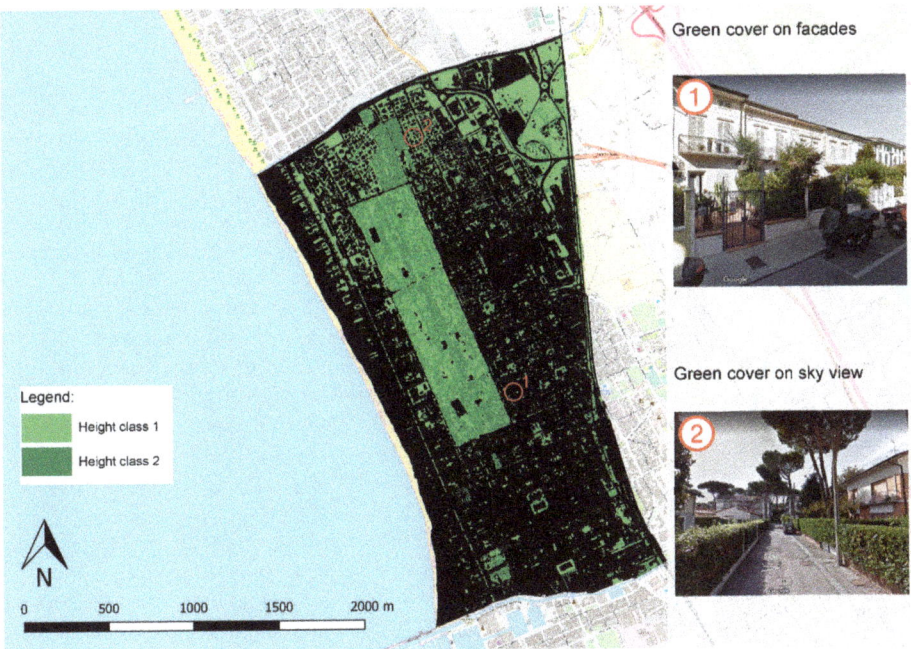

Figure 4. Urban green map classification.

The results were spatialized on the rasterized city blocks, as they are the central elements of urban planning and urban design on which the landscape metrics were calculated. As we considered that the ecological characteristics of urban greenery depend not only on the overall surface coverage, but also on the shape and distribution of vegetation within city blocks, we identified homogeneous city blocks through the use of landscape metrics. The metrics we used were the percentage of landscape (PLAND) and edge density (ED), which were normalized for the area. All metrics was calculated using

Fragstats 4.2 software. We chose PLAND because the city blocks had different dimensions, while ED is an important ecological parameter and many urban green benefits (such as pollution abatement, acting as an acoustic barrier, and being aesthetically pleasing) depend on the linear development and distribution of vegetation rather than its surface attributes [27–29].

Similar to the metrics derived from proximate sensing, the urban green metrics obtained from remote sensing were also calculated by classifying urban greenery into two classes through a map overlay operation between LiDAR and NDVI data. The two classes are: below the height of the block buildings and above the height of block buildings.

PLAND enables the percentage plant cover in each city block to be determined using the following calculation:

$$PLAND_i = \frac{\sum_{j=1}^{n} a_{ij}}{A} \qquad (5)$$

where $PLAND_i$ is the proportion of the landscape occupied by a patch type (class), i (below or over the height of buildings), a_{ij} is the area of patch i, j, and A is the total landscape area.

The ED_i metric enabled the compactness and distribution of the vegetation of each block to be determined using the following calculation:

$$ED_i = \sum_{k=1}^{m} e_{ik} \qquad (6)$$

where e_{ik} is the total length (m) of the landscape edge involving patch type (class), i, and includes the landscape boundary and background segments involving patch type i.

Spatial Clustering

As the ecological and visual characteristics of a city are manifested on a larger scale than that of a city block, it was necessary to create homogeneous areas by clustering city blocks based on their urban greenery characteristics.

As traditional clustering procedures do not consider the spatial relation between the geometries [30], we used a spatially bound geographic clustering procedure that grouped the territorial area objects into homogeneous contiguous regions [31].

Regionalization with dynamically constrained agglomerative clustering and partitioning (REDCAP) is a new method of spatial clustering and regionalization that was presented by Guo [32]. It is essentially based on a group of six methods for regionalization that are composed using a combination of three agglomerative clustering methods (single linkage clustering, SLK, average linkage clustering, AVG, and complete linkage clustering, CLK) and two different spatial constraining strategies: first-order constraining and full-order constraining [32]. The work of Guo [32] describes the technical and computational details of these six methods of regionalization, but we briefly describe here the theoretical context in which REDCAP is collocated, how it works in the case, and how it is applied in our analysis, of the AVG full-order method. The analysis is based on a contiguity matrix and a set of constrained strategies that drive the agglomerative clustering method. The average linkage clustering (ALK) defines the distance between two clusters as the average dissimilarity between all cross-cluster pairs of data points:

$$d_{ALK}(L, M) = \frac{1}{|L||M|} \sum_{u \in L} \sum_{u \in M} d_{uv} \qquad (7)$$

where $|L|$ and $|M|$ are the number of data points in clusters L and M, respectively, $u \in L$ and $v \in M$ are two data points, and d_{uv} is the dissimilarity between u and v.

The merging process incorporates the contiguity constraints using the full-order constraining strategy [32]. Contiguity-constrained agglomerative clustering requires that two clusters cannot be merged if they are not spatially contiguous, and this is the differential element between classic spatial clustering and regionalization. A full-order constraining strategy includes all edges in the clustering process, and the distance between two clusters is defined over all edges. This strategy is dynamic because it updates the contiguity matrix after each merge to track all edges which connect two different clusters.

3. Results

3.1. Image Segmentation

Using the Googleway library, we downloaded 14,542 geo-tagged images relating to 3638 sampling points acquired in 2018. Figure 5 shows the sampling points (Figure 5a) and some typical examples of the segmentation process (Figure 5b). The results shown in Table 1 show an overall accuracy of 89% and an accuracy of 83% for the tree classes.

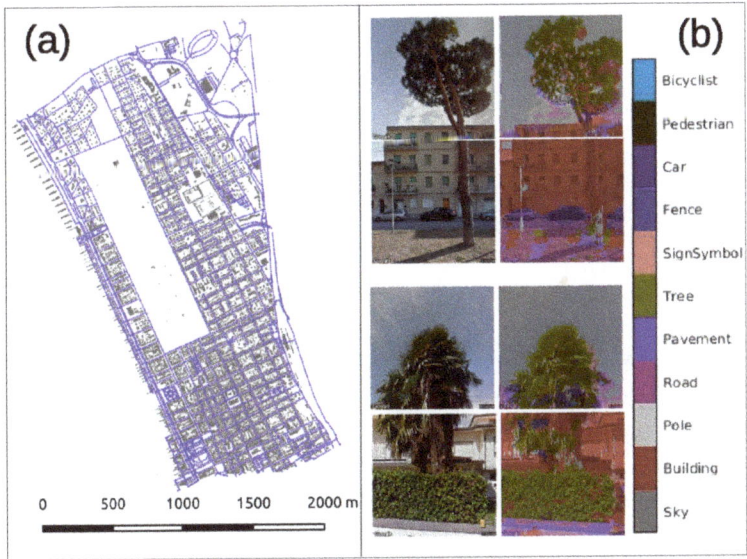

Figure 5. (a) Sampling points and (b) segmentation samples.

Table 1. Confusion matrix for network validation.

	Sky	Building	Pole	Road	Pavemnt	Tree	Sign.	Fence	Car	Pedestrian	Bicyclist.
Sky	0.94	0.01	0.02	0.00	0.00	0.04	0.00	0.00	0.00	0.00	0.00
Building	0.01	0.80	0.07	0.00	0.01	0.02	0.05	0.02	0.01	0.01	0.00
Pole	0.01	0.07	0.77	0.00	0.01	0.03	0.05	0.03	0.01	0.02	0.00
Road	0.00	0.00	0.00	0.94	0.04	0.00	0.00	0.00	0.01	0.00	0.00
Pavement	0.00	0.01	0.01	0.02	0.93	0.00	0.00	0.01	0.01	0.01	0.00
Tree	0.02	0.02	0.03	0.00	0.00	0.88	0.01	0.03	0.00	0.00	0.00
SignSymbol	0.00	0.04	0.05	0.00	0.00	0.01	0.88	0.01	0.01	0.00	0.00
Fence	0.00	0.01	0.02	0.00	0.00	0.01	0.00	0.93	0.01	0.01	0.00
Car	0.00	0.00	0.00	0.01	0.00	0.00	0.00	0.01	0.02	0.91	0.03	0.01
Pedestrian	0.00	0.02	0.02	0.00	0.01	0.00	0.00	0.01	0.02	0.90	0.01
Bicyclist	0.00	0.00	0.01	0.00	0.01	0.00	0.00	0.00	0.01	0.02	0.95

Figure 6 shows the receiver operation curves (ROC) curves and the values of the area under the ROC curve (AUC) for the different classes calculated by means of a sub-sample of 50 images. The area under the curve for all the four class is very large indicating the high accuracy of the algorithm.

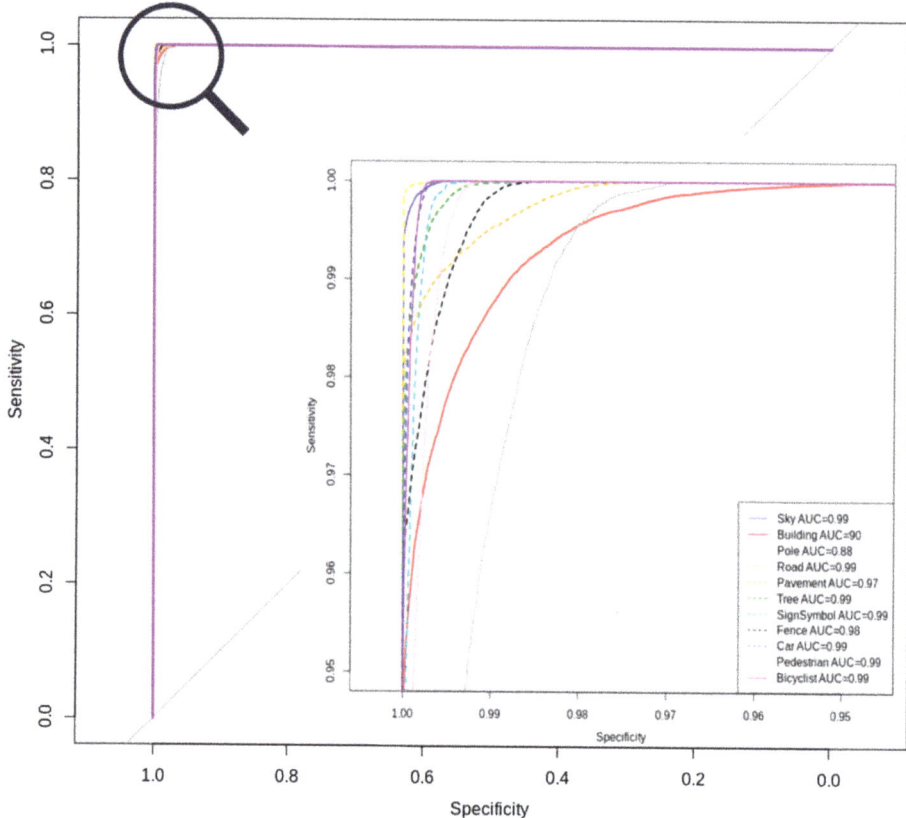

Figure 6. Left: receiver operation characteristics (ROC) curves obtained for DeepLabV3+ architecture for semantic segmentation. Zoomed view of the selected area is shows in the right figure. Area under the curve (AUC) values for each method is reported in the legend.

These results are in line with those obtained in other researches that have used DeepLabV3+ in remote sensing [33] or in other fields of study [34]. Few segmentation outputs obtained using DeepLabV3+ are in the supplementary materials. It can be seen that the green class is segmented accurately with a sharp class boundary.

The Figure 7a shows the frequency distribution of the two metrics derived from segmentation of street view images. The city of Viareggio is characterized by greenery affecting most of the lower facades of buildings on public roads. The GCF index has an average value of 10% (median = 8.6%) with the first quartile of 2.9% and the third quartile of 14.6%. However, the GCS index has lower values, with an average of 3.1% (median 1.3%) and first and third quartiles of 0.2% and 4.3%, respectively.

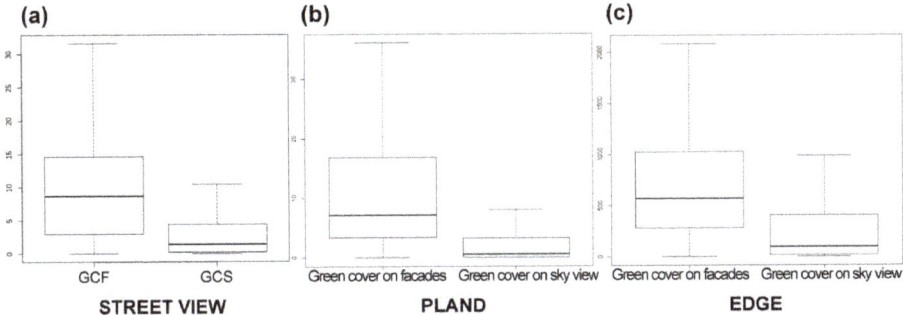

Figure 7. Frequency distribution metrics: (**a**) street view metrics (**b**) percentage of landscape (PLAND) and (**c**) edge density (EDGE).

We used the network-based inverse distance weighting (NT-IDW) [35] method to spatialize the relative sampling point values in raster maps of the two indices. The NT-IDW method expands the commonly used spatial interpolation methods, IDW (inverse distance weighting) and inverse distance weighting, and the results are applied to analyze spatial data observed on a network. IDW assumes that all locations exist in a two-dimensional Euclidean space, (the distance between the sample locations and the target locations are measured using the straight-line distance). NT-IDW extends from IDW but uses a network distance instead. NT-IDW was conducted using the routines in the ipdw R package [36], and Figure 8 shows the results of spatialization using the NT-IDW method. The GCF index has higher values in the northern part of the city, especially in the north-west quadrant, where the so-called "garden city" is located. In contrast, the GCS index has relatively high values within isolated hot spots scattered across the entire urban area.

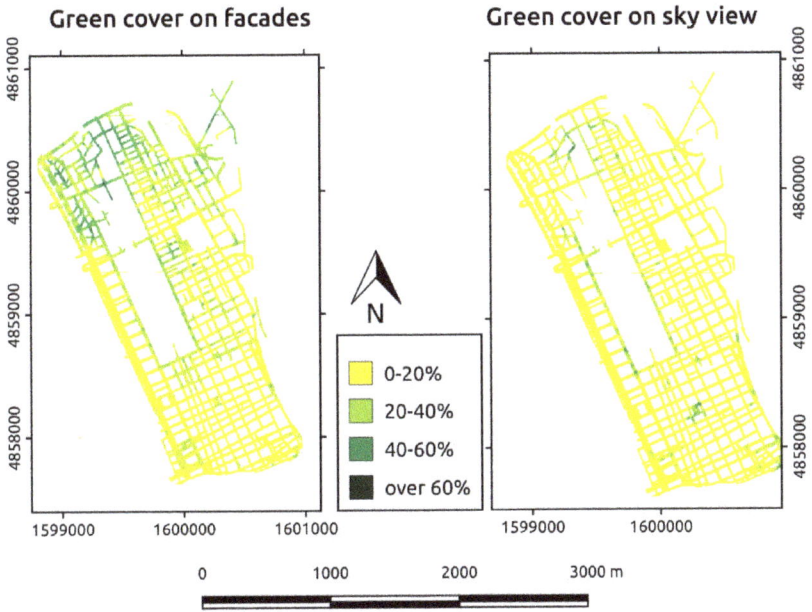

Figure 8. Maps of Grenn Cover on Facades (GCF) and Green Cover on Sky (GCS) indices.

3.2. Remote Sensing Landscape Metrics

The Figures 7b,c show box and whisker plots of landscape metrics distribution of the green cover on facades compared to the green cover on sky view for PLAND and ED, respectively.

The results of PLAND (Figure 7b) show the percentage of green cover and also show that most of the city has green cover below the height of the buildings rather than higher than buildings. The former has a first quartile value of 3%, median value of 7%, and third quartile value of 17%, while the latter has a first quartile value of 0%, median value of 0.5%, and third quartile value of 3%.

The results of ED (Figure 7c) show the no-compactness values of vegetation, which are negatively related to the compactness of urban green. An analysis of the boxplot shows that the city is mostly has green cover below the height of buildings (first quartile value of 280 m/ha, median value of 567 m/ha, third quartile value of 1028 m/ha) but less green cover over the height of buildings (first quartile value of 16 m/ha, median value of 94.5 m/ha, and third quartile value of 405 m/ha).

These results are visually represented through two maps that show the quantile distribution of different vegetation types: green cover on facades and on sky view (Figure 9). In both, the green cover below the height of buildings is predominantly located north of the study area, and its distribution is more compact than that of sky view, which is mainly located south of the city and is more fragmented.

In order to evaluate, the homogeneity of the metrics within each block were calculated by the standard deviation statistics within and between the blocks. In fact, the verification of the homogeneity of the metrics within the blocks allows to verify the efficiency of the blocks as a tessellation element of the study area. The results shown in Table 2 show that the blocks are an efficient tessellation of the urban space. The descriptive statistics of the 351 isolates are available as supplementary material.

Table 2. Statistics of metrics within and between the blocks.

Variable	Within block Mean Square (a)	Between block Mean Square (b)	Ratio a/b
Edge density on facades	10.7067	1050.1234	0.0102
Edge density on Sky	9.7539	622.3140	0.0157
Percent of landscape on facades	0.0030	0.4060	0.0075
Percent of landscape on Sky	0.0016	0.0835	0.0188
Green Cover on facades	1.2533×10^{-5}	3.5×10^{-6}	3.5792
Green Cover on Sky view	7.1352×10^{-6}	3.5016×10^{-6}	2.0377
Number of pixel Landscape metrics	3,553,123		
Degree of freedom landscape metrics	3,552,772		
N. of pixel Street View metrics	58,218		
Degree of freedom Street View metrics	57,867		
Degree of freedom Blocks	351		

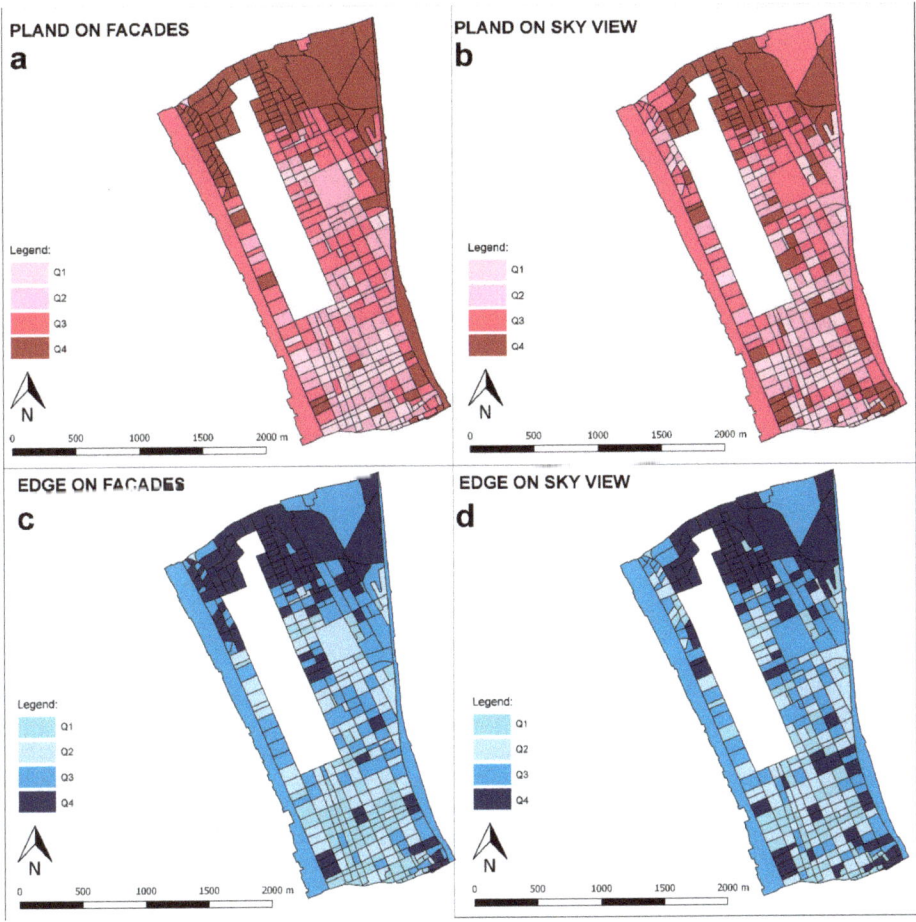

Figure 9. Maps of remote sensing metrics: (**a**) percent of landscape on facades; (**b**) percent of landscape on Sky; (**c**) edge density on facades; (**d**) edge density on Sky.

3.3. Results of Spatial Clustering with REDCAP Method

As it is necessary to create homogeneous areas by clustering city blocks, Pakzad and Salari [37] explained the ecological and visual characteristics of the city on a larger scale than that of the city block. Therefore, we applied REDCAP as the regionalization tool, which enabled the city to be divided into homogeneous and spatially close clusters based on the same green cover characteristics. The metrics used were: PLAND over block height (PLANDOvB) and PLAND below block height (PLANDBelB), ED over block height (EdgeOvB) and ED below block height (EdgeBelB), and street view on facades (GCF) and street view on sky view (GCS).

Figure 10 shows the correlation between metrics used.

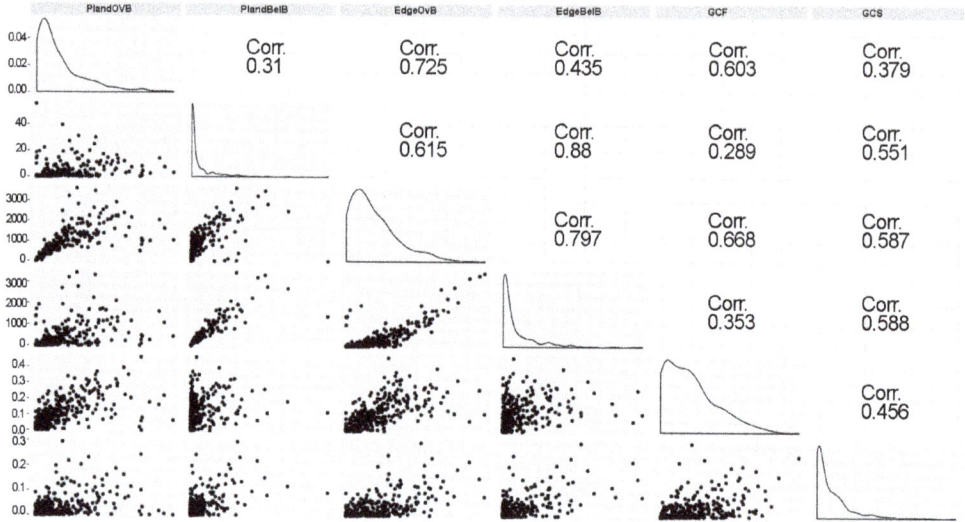

Figure 10. Correlation matrix. Main diagonal graph are the frequency distribution of variables. Cells above the main diagonal show the correlation coefficients. Graph above the main diagonal is the scatter plot of the variables.

The regionalization approach is sensitive to correlated attributes during the calculation of homogeneity between regions, and many of the attributes are derived from the same data, leading to a correlation between attributes. Thus, it was necessary to apply principal component analysis (PCA) to the variable's metrics, and unique information could then be retained while excluding correlated information between variables [38–40]. We chose the first three dimensions based on the 95% threshold criterion [41]. Table 3 shows the loading variable of the first three PCs. The first component loads positively on ED metrics, the second PC loads negatively on the metrics below the level of the isolates (GCF, PLANDBelB) and positively with PLANDOvB, and the third PC has a high GCS load.

Table 3. Principal component analysis (PCA) dimensions.

	PC1	PC2	PC3
Green Cover on Facades	0.3559	−0.5382	0.2702
Green Cover on Sky	0.3834	0.1345	0.8119
Percent of landscape below block height	0.3686	−0.4999	−0.3795
Percent of landscape over block height	0.4011	0.5186	−0.1583
Edge density below block height	0.4789	−0.1277	−0.1980
Edge denisty over block height	0.4475	0.3964	−0.2440
% of Variance	63.0674	23.8025	9.1246

We therefore used the values of PC1, PC2, and PC3 calculated for each of the 351 isolates as input data to identify clusters of contiguous and homogeneous blocks through the REDCAP method. Since the REDCAP method requires to specify a priori the number of clusters to be created, it was necessary to find the optimal number of clusters. We selected the elbow method to determine the optimal number of clusters, as this method optimizes the variance within clusters [42]. This method looks at the variance within the clusters as a function of the number of clusters: One should choose a number of clusters so that adding another cluster does not give much better modeling of the data. More precisely, if one plots the variance within the clusters against the number of clusters, the first clusters will add much information (explain a lot of variance), but at some point the marginal gain will drop, giving an angle in the graph. The number of clusters is chosen at this point, hence the "elbow criterion". The elbow diagram in Figure 10 shows that when there are more than 10 clusters, there is no significant decrease in the variance within the clusters. So, the optimal number of clusters was 10.

Figures 11b and 12a respectively, show the frequency distribution of ecological metrics within the clusters of homogeneous blocks identified through the REDCAP procedure. Cluster values are shown in Table 4. The lowest value for all metrics is visible in cluster 2, which includes only built city blocks. Furthermore, clusters 1 and 10 are urban blocks with a low presence of vegetation; although they are similar to each other, they belong to two different clusters because they are not contiguous. This characteristic is typical of the geographically constrained clustering methodology REDCAP, which can identify clusters of similar blocks in the metric values as long as they are distant in the geographical space. The maximum values, which are representative of a greater presence and distribution of vegetation, are identified in clusters 9, 7, and 3. These clusters are within the northern area, where there is a greater percentage of greenery and where hedges and rows of large trees are higher than the facades of buildings.

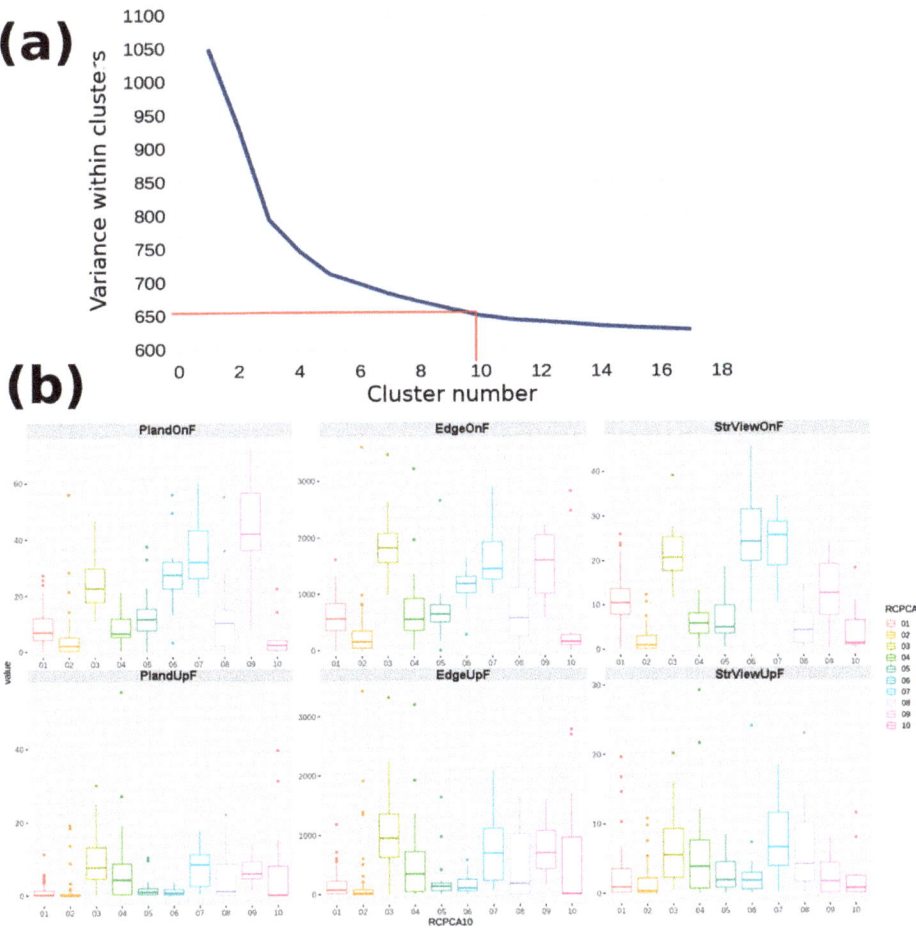

Figure 11. Spatial clustering results. (**a**) Elbow method and choice of the optimal number of clusters. (**b**) Frequency distribution boxplot of the green metrics of the blocks within each cluster.

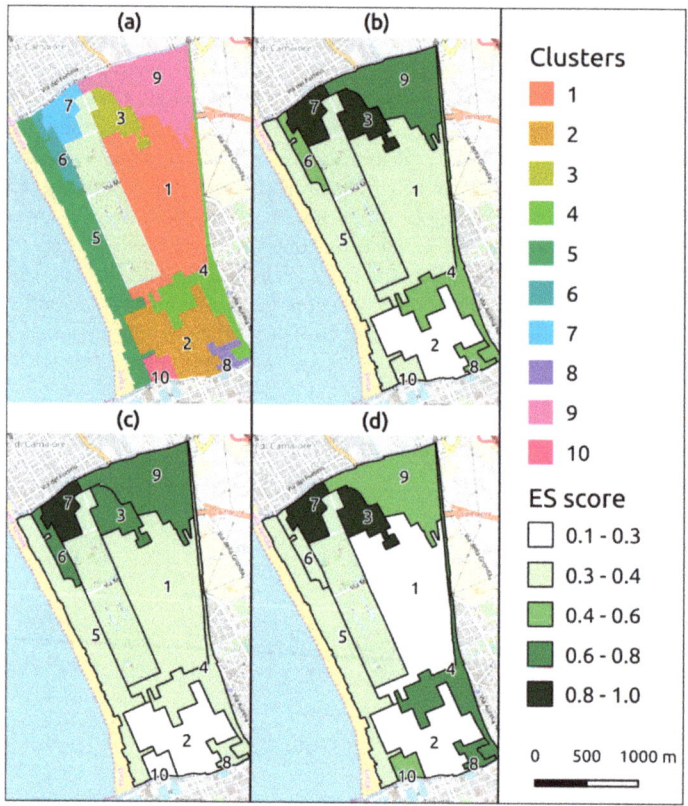

Figure 12. (**a**) Cluster map, (**b**) Ecosystem Services total score map, (**c**) Ecosystem Services score below block heights, and (**d**) Ecosystem Services score above the block height.

To summarize the results obtained from the 6 metrics, we proposed the use of three Ecosystem Services (ES) score indices: ES score below block heights (ES^k_{below}), ES score above block heights (ES^k_{above}), and ES total score (ES^k_{tot}), which are formulated as follows:

$$ES^k_{below} = \frac{\sum \frac{PlandBelB_k}{max_k(PlandBelB_k)} + \frac{EdgeBelB_k}{max_k(EdgeBelB_k)} + \frac{GCF_k}{max_k(GCF_k)}}{3} \tag{8}$$

$$ES^k_{above} = \frac{\sum \frac{PlandOverB_k}{max_k(PlandOverB_k)} + \frac{EdgeOverB_k}{max_k(EdgeOverB_k)} + \frac{GCS_k}{max_k(GCS_k)}}{3} \tag{9}$$

$$ES^k_{tot} = \frac{ES^k_{below} + ES^k_{above}}{2} \tag{10}$$

Figure 12b–d shows maps of the three ES indices.

Table 4. Mean, median, first, and tirth quartiles of metrics of the blocks within each cluster. PLANDBelB and PLANDOvB are percent of green detected by remote sensing respectively, below and over the height of block. EdgeBelB and EdgeOvB are the edge density of green detected by remote sensing respectively, below and over the height of block. GCF and GCS are the percent of green detected by segmentation of GSV images respectively, on facades and on sky view.

		Clusters									
		1	2	3	4	5	6	7	8	9	10
PLANDBelB	Mean	8.3	3.9	24.4	8.4	13.0	27.4	34.5	13.4	45.6	4.6
	Q1	4.3	0.4	17.6	5.2	7.4	22.5	26.2	2.1	36.1	0.4
	Median	6.9	2.1	22.6	6.4	11.4	27.4	31.8	10.1	41.8	2.1
	Q3	12.1	5.1	29.7	11.7	15.1	32.0	43.2	14.8	56.4	3.8
PLANDOvB	Mean	1.1	1.2	9.5	7.3	1.9	1.0	8.1	5.9	6.6	7.9
	Q1	0.1	0.0	4.6	0.3	0.3	0.2	2.5	0.0	4.5	0.0
	Median	0.3	0.1	7.7	4.3	0.9	0.6	8.5	1.1	5.9	0.0
	Q3	1.4	0.5	13.2	8.7	1.9	1.6	11.2	8.7	9.2	7.9
EdgeBelB	Mean	608.8	253.1	1884.1	768.8	706.7	1146.3	1608.3	769.5	1493.7	593.9
	Q1	362.5	48.2	1566.5	355.9	505.7	1020.4	1262.2	252.0	1006.3	93.3
	Median	565.6	157.8	1824.9	554.0	646.5	1183.5	1449.2	574.8	1596.8	150.5
	Q3	838.2	349.0	2073.6	919.6	807.7	1311.6	1924.1	1099.8	2041.3	266.3
EdgeOvB	Mean	152.0	150.2	1053.5	540.3	238.0	165.3	720.9	498.2	769.2	672.8
	Q1	18.0	0.0	631.2	50.6	64.1	59.0	236.0	0.0	426.9	0.0
	Median	80.0	21.5	950.4	344.2	134.4	105.7	693.9	180.5	692.3	6.3
	Q3	226.9	80.4	1354.7	724.0	187.5	252.8	1111.5	1021.9	1071.1	956.3
GCF	Mean	11.1	2.0	21.5	6.0	7.3	26.0	23.6	5.1	12.9	4.6
	Q1	8.0	0.2	17.7	3.6	3.7	19.9	18.8	1.5	7.6	1.0
	Median	10.5	1.0	20.7	5.9	5.0	24.3	25.8	4.3	12.6	1.3
	Q3	13.7	3.1	25.3	8.1	10.0	31.5	28.7	8.0	19.1	6.4
GCS	Mean	2.2	1.4	6.8	5.6	2.7	3.1	8.1	6.7	2.6	2.4
	Q1	0.2	0.1	2.3	0.7	0.9	0.6	3.9	1.6	0.1	0.1
	Median	0.9	0.4	5.5	3.9	1.9	1.9	6.6	4.1	1.7	0.7
	Q3	3.6	2.2	9.3	7.7	4.5	3.0	11.6	10.1	4.3	2.4

The figures show that areas with the lowest scores for the three ES indices are clusters 1, 2, and 5. The scores are particularly low in cluster 5 (see Figure 13a), which includes the promenade and is the most important tourist area of the city. Clusters 3, 6, 7, and 9 have the highest scores for all three indices and are representative of the so-called "garden city" which was recently developed. The high values for the above-block height scores occur in relation to plants preserved in the pine forest within the urban area (Figure 13b). Clusters 1, 2, 4, 8, and 10 are all characterized by a high building density, and slightly different scores are obtained. For example, there are rows of small plants along the sidewalks (Figure 13c) or trees in the middle of roads (Figure 13d) in clusters 4 and 8, whereas clusters 1, 2, and 10 have almost no public greenery, despite having a similar urban fabric.

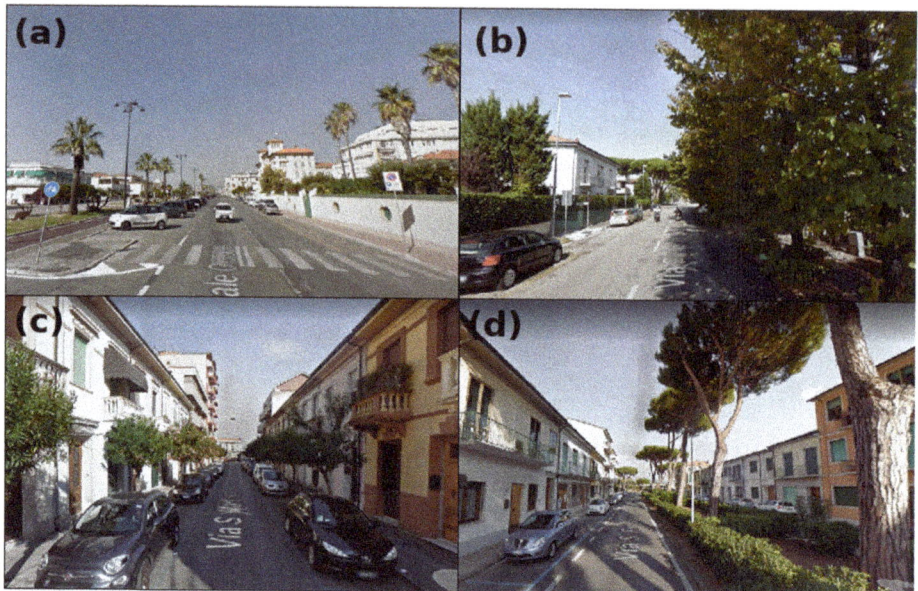

Figure 13. Typical representative public greenery: (**a**) promenade, (**b**) pine forest within the urban area, (**c**) are rows of small plants along the sidewalks, (**d**) trees in the middle of roads.

4. Discussion

This paper proposes a set of ecological metrics for urban greenery that can be used to evaluate urban green ecosystem services. The metrics were derived from two complementary surveys: a traditional remote sensing survey using multispectral images and LiDAR data, and a proximate sensing survey using images available on the GSV database.

With respect to the classification of vegetated areas, the results of our study confirm the efficacy of the combined use of NDVI and LiDAR data for remotely-sensed images [43,44] and of semantic segmentation using deep learning for GSV images [45,46].

The methodology is unique, as it identifies metrics (based on existing literature) of urban green ecosystem services using both data sources. The ecosystem services of urban forests were derived from greenery shading facades of buildings and from the covering of roads, roofs, and courtyards by the foliage of trees. The metrics were then divided into urban greenery below the height of buildings and that above the height of buildings. The correlation matrix in Figure 10 shows that the proximate sensing metrics agree with those derived from remote sensing data. The partial correlation (about 0.6) between GCF/GCS and the PLAND/ED indices shows that both reliefs can map the presence of vegetation with good coherence, but they reveal different characteristics. Proximate sensing data more efficiently highlights the shading of facades and streets, but the data are limited to public spaces that are accessible via a vehicle. In contrast, the data from remote sensing are more efficient for highlighting roofing and the shading of private courtyards.

The urban green survey using proximate sensing images at street level can be conducted at a low cost and can thus be used to monitor the health status of urban green spaces. However, further research is required to define methodologies (based on the use of commercial spherical cameras integrated with medium-scale satellite images) that can be used by city administrations (Landsat and Sentinel) [45,46] to conduct low-cost, short-interval monitoring.

In accordance with previous studies [47–50], we report a set of metrics for city blocks. The research also made it possible to create a methodology for verifying the ecological homogeneity of urban blocks which can therefore be considered efficient tessellation units for managing urban green areas.

Indeed Sheppard et al. [51] argue that the city block scale is an often neglected but promising level for community engagement and co-creation of climate change responses.

However, an innovative element of this study is the application of regionalization techniques through constrained clustering that are used to identify homogeneous areas, with the aim of determining the horizontal and vertical characteristics of urban greenery. This method will enable urban planners to identify areas within a city that can be considered greener than others, and it can also be used as a monitoring tool when conducting a differentiated analysis of gain or loss with respect to urban street greenery. In addition, it could be used in the planning stage of an urban greening program to assist urban planners in selecting appropriate locations, sizes, and types of greenery that provide the maximum affect. Furthermore, it could be employed to check the visual impact of urban forest management practices and document the visibility of urban greenery in cities. The work also showed that the block is a sufficiently homogeneous and therefore efficient geographical unit for the classification of urban green. A finer geographical unit could have the advantage of a greater homogeneity of the metrics, but it would make it more difficult to transfer the results of the research to guide management and improvement interventions of the green ecosystem services in the city.

The block clusters obtained with the REDCAP method in fact meet the requirements set by Barron et al. [52] (p. 21) to define the "neighborhood scale" as a set of cohesive blocks. They highlight that "The neighborhood scale captures human green space experiences at shorter distances, allowing for consideration of accessibility, sightlines, aesthetics, vegetation layering, and quality of greenspace design. The experiential neighborhood scale is small enough to conceptually include the impact of individual trees, an important component of urban greens paces". Therefore, the authors in their work propose interventions that provide strategic green space enhancement at the neighborhood and block scale.

The high diversification and complementarity of the metrics proposed at the city block level enable a better understanding of the role that trees and vegetation play in urban dynamics and human health. Although some studies have shown a link between human health benefits and the presence of urban greenery [53–55], these studies have used small samples and a limited geographical scope for a single route. The ability to measure a complete set of ecological green urban metrics would enable researchers to determine whether the health benefits of urban trees are pervasive, whether they exist in specific contexts with respect to biogeographical conditions, or whether they can be maximized (for example with respect to local policies, management practices, and socioeconomic indicators). This method enables urban tree cover and its relationship with local conditions or social factors to be analyzed on a much finer scale than previously possible. The relationships between urban trees and physical and social components of cities that were previously opaque, such as how income level and social status relate to tree presence and neighborhood aesthetics, can now be investigated in depth.

5. Conclusions

Our study proposes a set of ecological green urban metrics based on the integration of remote sensing and proximate sensing data. Metrics derived from proximate sensing were calculated by classifying the urban forest using GSV database images. Green areas were classified by semantic segmentation using pretrained deep neural networks. The results demonstrate the high efficiency of the method, which has an accuracy of 83%.

Metrics from remote sensing were derived from overlaying multi-spectral images and LiDAR data. In accordance with previous studies, two classes of metrics were calculated: greenery at a lower or higher elevation than building facades, respectively.

In the last phase of the work, the metrics were related to city blocks, and homogeneous areas were identified relative to the characteristics of urban green using a spatially constrained clustering methodology.

The proposed methodology enables the creation of a geographic information system that can be used by public administrators and urban green designers to maintain and create new public forests in cities.

Furthermore, research can be further developed in three possible ways: by expanding the survey to other urban elements, in particular buildings and roads, by verifying the relationships between homogeneous clusters that show characteristics of landscape ecology and the visual quality of the city, climatic well-being, or as a pollution reduction strategy, or by employing different and more complex landscape metrics.

Supplementary Materials: The following are available online at http://www.mdpi.com/2072-4292/12/2/329/s1. grassProcedure.txt: GRASS procedure. MATLAB_procedure.m: MATLAB procedure. R_procedure.R: Cran R procedure. SegExamp.zip: Sample of segmented Google Street View images. CityBlock.xlsx: descriptive statistics of city blocks metrics.

Author Contributions: Conceptualization, I.B., I.C., and E.B.; methodology, I.B. and E.B.; software, I.B. and E.B.; validation, I.B. and C.S.; writing—original draft preparation, I.B.; writing—review and editing, E.B. and I.B.; visualization, E.B. and I.C.; supervision, C.S.; funding acquisition, C.S. The manuscript was written by I.B. and improved by the contributions of all the co-authors. All authors have read and agreed to the published version of the manuscript.

Funding: The authors acknowledge financial support from the "Unione dei comuni Circondario dell'Empolese Valdelsa".

Conflicts of Interest: The authors declare no conflicts of interest.

References

1. Salmond, J.A.; Tadaki, M.; Vardoulakis, S.; Arbuthnott, K.; Coutts, A.; Demuzere, M.; McInnes, R.N. Health and climate related ecosystem services provided by street trees in the urban environment. *Environ. Health* **2016**, *15*, 95–111. [CrossRef] [PubMed]
2. Mavrogianni, A.; Davies, M.; Taylor, J.; Chalabi, Z.; Biddulph, P.; Oikonomou, E.; Jones, B. The impact of occupancy patterns, occupant-controlled ventilation and shading on indoor overheating risk in domestic environments. *Build. Environ.* **2014**, *78*, 183–198. [CrossRef]
3. Berry, R.; Livesley, S.J.; Aye, L. Tree canopy shade impacts on solar irradiance received by building walls and their surface temperature. *Build. Environ.* **2013**, *69*, 91–100. [CrossRef]
4. Streiling, S.; Matzarakis, A. Influence of single and small clusters of trees on the bioclimate of a city: A case study. *J. Arboric.* **2003**, *29*, 309–316.
5. Oke, T.R. The micrometeorology of the urban forest. *Philos. Trans. R. Soc. Lond. B Biol. Sci.* **1989**, *324*, 335–349. [CrossRef]
6. Gorbachevskaya, O.; Schreiter, H.; Kappis, C. Wissenschaftlicher Erkenntnisstand über das Feinstaubfilterungspotential von Pflanzen (qualitativ und quantitativ). Ergebnisse der Literaturstudie. *Berl. Geogr. Arb.* **2007**, *109*, 71–82.
7. Litschke, T.; Kuttler, W. On the reduction of urban particle concentration by vegetation—A review. *Meteorol. Z.* **2008**, *17*, 229–240. [CrossRef]
8. Nilsson, M.; Bengtsson, J.; Klæboe, R. (Eds.) *Environmental Methods for Transport Noise Reduction*; CRC Press: Boca Raton, FL, USA, 2014.
9. Jiang, B.; Li, D.; Larsen, L.; Sullivan, W.C. A dose-response curve describing the relationship between urban tree cover density and self-reported stress recovery. *Environ. Behav.* **2016**, *48*, 607–629. [CrossRef]
10. Leung, D.; Newsam, S. Proximate sensing: Inferring what-is-where from georeferenced photo collections. In Proceedings of the 2010 IEEE Computer Society Conference on Computer Vision and Pattern Recognition, San Francisco, CA, USA, 13–18 June 2010; pp. 2955–2962.
11. Homer, C.; Dewitz, J.; Fry, J.; Coan, M.; Hossain, N.; Larson, C.; Wickham, J. Completion of the 2001 national land cover database for the conterminous United States. *Photogramm. Eng. Remote Sens.* **2007**, *73*, 337.
12. Barbierato, E.; Bernetti, I.; Capecchi, I.; Saragosa, C. Remote sensing and urban metrics: An automatic classification of spatial configurations to support urban policies. In *Earth Observation Advancements in a Changing World*; Italian Society of Remote Sensing: Torino, Italy, 2019; p. 187.

13. Kardan, O.; Gozdyra, P.; Misic, B.; Moola, F.; Palmer, L.J. Neighborhood greenspace and health in a large urban center. In *Urban Forests*; Apple Academic Press: New York, NY, USA, 2017; pp. 77–108.
14. Li, X.; Zhang, C.; Li, W.; Ricard, R.; Meng, Q.; Zhang, W. Assessing street-level urban greenery using Google Street View and a modified green view index. *Urban For. Urban Green.* **2015**, *14*, 675–685. [CrossRef]
15. Li, X.; Ratti, C.; Seiferling, I. Quantifying the shade provision of street trees in urban landscape: A case study in Boston, USA, using Google Street View. *Landsc. Urban Plan.* **2018**, *169*, 81–91. [CrossRef]
16. Seiferling, I.; Naik, N.; Ratti, C.; Proulx, R. Green streets–Quantifying and mapping urban trees with street-level imagery and computer vision. *Landsc. Urban Plan.* **2017**, *165*, 93–101. [CrossRef]
17. Li, X.; Zhang, C.; Li, W.; Kuzovkina, Y.A. Environmental inequities in terms of different types of urban greenery in Hartford, Connecticut. *Urban For. Urban Green.* **2016**, *18*, 163–172. [CrossRef]
18. Middel, A.; Lukasczyk, J.; Zakrzewski, S.; Arnold, M.; Maciejewski, R. Urban form and composition of street canyons: A human-centric big data and deep learning approach. *Landsc. Urban Plan.* **2019**, *183*, 122–132. [CrossRef]
19. Cao, R.; Zhu, J.; Tu, W.; Li, Q.; Cao, J.; Liu, B.; Zhang, Q.; Qiu, G. Integrating Aerial and Street View Images for Urban Land Use Classification. *Remote Sens.* **2018**, *10*, 1553. [CrossRef]
20. Hermosilla, T.; Palomar-Vázquez, J.; Balaguer-Beser, Á.; Balsa-Barreiro, J.; Ruiz, L.A. Using street based metrics to characterize urban typologies. *Comput. Environ. Urban Syst.* **2014**, *44*, 68–79. [CrossRef]
21. Leung, D.; Newsam, S. Can off-the-shelf object detectors be used to extract geographic information from geo-referenced social multimedia? In Proceedings of the 5th ACM SIGSPATIAL International Workshop on Location-Based Social Networks, Redondo Beach, CA, USA, 7–9 November 2012; ACM: New York, NY, USA, 2012; pp. 12–15.
22. MathWorks. Semantic Segmentation Using Deep Learning. Available online: https://it.mathworks.com/help/vision/examples/semantic-segmentation-using-deep-learning.html (accessed on 1 September 2019).
23. Chen, L.C.; Zhu, Y.; Papandreou, G.; Schroff, F.; Adam, H. Encoder-Decoder with Atrous Separable Convolution for Semantic Image Segmentation. In Proceedings of the ECCV, Munich, Germany, 8–14 September 2018.
24. Brostow, G.J.; Fauqueur, J.; Cipolla, R. Semantic object classes in video: A high-definition ground truth database. *Pattern Recognit. Lett.* **2009**, *30*, 88–97. [CrossRef]
25. Rodgers, J.C.; Murrah, A.W.; Cooke, W.H. The Impact of Hurricane Katrina on the Coastal Vegetation of the Weeks Bay Reserve, Alabama from NDVI Data. *Estuaries Coasts* **2009**, *32*, 496–507. [CrossRef]
26. McGarigal, K.; Cushman, S.A.; Ene, E. FRAGSTATS v4: Spatial Pattern Analysis Program for Categorical and Continuous Maps. Computer Software Program Produced by the Authors at the University of Massachusetts, Amherst. 2012. Available online: http://www.umass.edu/landeco/research/fragstats/fragstats.html (accessed on 16 January 2020).
27. Maimaitiyiming, M.; Ghulam, A.; Tiyip, T.; Pla, F.; Latorre-Carmona, P.; Halik, Ü.; Sawut, M.; Caetano, M. Effects of green space spatial pattern on land surface temperature: Implications for sustainable urban planning and climate change adaptation. *ISPRS J. Photogramm. Remote Sens.* **2014**, *89*, 59–66. [CrossRef]
28. Tian, Y.; Jim, C.Y.; Wang, H. Assessing the landscape and ecological quality of urban green spaces in a compact city. *Landsc. Urban Plan.* **2014**, *121*, 97–108. [CrossRef]
29. Cho, S.H.; Poudyal, N.C.; Roberts, R.K. Spatial analysis of the amenity value of green open space. *Ecol. Econ.* **2008**, *66*, 403–416. [CrossRef]
30. MacQueen, J. Some methods for classification and analysis of multivariate observations. In Proceedings of the Fifth Berkeley Symposium on Mathematical Statistics and Probability, Berkeley, CA, USA, 21 June 1967; 1: Statistics. University of California Press: Berkeley, CA, USA, 1967; pp. 281–297.
31. Haining, R. *Spatial Data Analysis: Theory and Practice*; Cambridge University Press: Cambridge, UK, 2003.
32. Guo, D. Regionalization with dynamically constrained agglomerative clustering and partitioning (REDCAP). *Int. J. Geogr. Inf. Sci.* **2008**, *22*, 801–823. [CrossRef]
33. Girisha, S.; Pai, M.M.; Verma, U.; Pai, R.M. Performance Analysis of Semantic Segmentation Algorithms for Finely Annotated New UAV Aerial Video Dataset (ManipalUAVid). *IEEE Access* **2019**, *7*, 136239–136253. [CrossRef]
34. Jahanifar, M.; Tajeddin, N.Z.; Koohbanani, N.A.; Gooya, A.; Rajpoot, N. Segmentation of Skin Lesions and their Attributes Using Multi-Scale Convolutional Neural Networks and Domain Specific Augmentations. *arXiv* **2018**, arXiv:1809.10243.

35. Shiode, N.; Shiode, S. Street-level spatial interpolation using network-based IDW and ordinary kriging. *Trans. GIS* **2011**, *15*, 457–477. [CrossRef]
36. Stachelek, J. Ipdw: Interpolation by Inverse Path Distance Weighting. R Package Version 0.2-1. 2014. Available online: http://CRAN.R-project.org/package=ipdw (accessed on 16 January 2020).
37. Pakzad, E.; Salari, N. Measuring sustainability of urban blocks: The case of Dowlatabad, Kermanshah city. *Cities* **2018**, *75*, 90–100. [CrossRef]
38. Adams, M.D.; Kanaroglou, P.S.; Coulibaly, P. Spatially constrained clustering of ecological units to facilitate the design of integrated water monitoring networks in the St. Lawrence Basin. *Int. J. Geogr. Inf. Sci.* **2016**, *30*, 390–404. [CrossRef]
39. Helbich, M.; Brunauer, W.; Hagenauer, J.; Leitner, M. Data-driven regionalization of housing markets. *Ann. Assoc. Am. Geogr.* **2013**, *103*, 871–889. [CrossRef]
40. Wang, F.; Guo, D.; McLafferty, S. Constructing geographic areas for cancer data analysis: A case study on late-stage breast cancer risk in Illinois. *Appl. Geogr.* **2012**, *35*, 1–11. [CrossRef] [PubMed]
41. Anselin, L.; Syabri, I.; Kho, Y. GeoDa: An introduction to spatial data analysis. *Geogr. Anal.* **2006**, *38*, 5–22. [CrossRef]
42. Kassambara, A. *Practical Guide to Cluster Analysis in R: Unsupervised Machine Learning*; STHDA: Montpellier, France, 2017; Volume 1.
43. Zhang, C.; Zhou, Y.; Qiu, F. Individual tree segmentation from LiDAR point clouds for urban forest inventory. *Remote Sens.* **2015**, *7*, 7892–7913. [CrossRef]
44. Huang, Y.; Yu, B.; Zhou, J.; Hu, C.; Tan, W.; Hu, Z.; Wu, J. Toward automatic estimation of urban green volume using airborne LiDAR data and high resolution Remote Sensing images. *Front. Earth Sci.* **2013**, *7*, 43–54. [CrossRef]
45. Stubbings, P.; Peskett, J.; Rowe, F.; Arribas-Bel, D. A Hierarchical Urban Forest Index Using Street-Level Imagery and Deep Learning. *Remote Sens.* **2019**, *11*, 1395. [CrossRef]
46. Zhou, H.; He, S.; Cai, Y.; Wang, M.; Su, S. Social inequalities in neighborhood visual walkability: Using Street View imagery and deep learning technologies to facilitate healthy city planning. *Sustain. Cities Soc.* **2019**, 101605. [CrossRef]
47. Fu, Y.; Li, J.; Weng, Q.; Zheng, Q.; Li, L.; Dai, S.; Guo, B. Characterizing the spatial pattern of annual urban growth by using time series Landsat imagery. *Sci. Total Environ.* **2019**, *666*, 274–284. [CrossRef]
48. Deng, J.; Huang, Y.; Chen, B.; Tong, C.; Liu, P.; Wang, H.; Hong, Y. A Methodology to Monitor Urban Expansion and Green Space Change Using a Time Series of Multi-Sensor SPOT and Sentinel-2A Images. *Remote Sens.* **2019**, *11*, 1230. [CrossRef]
49. Voltersen, M.; Berger, C.; Hese, S.; Schmullius, C. Object-based land cover mapping and comprehensive feature calculation for an automated derivation of urban structure types at block level. *Remote Sens. Environ.* **2014**, *154*, 192–201. [CrossRef]
50. Vanderhaegen, S.; Canters, F. Mapping urban form and function at city block level using spatial metrics. *Landsc. Urban Plan.* **2017**, *167*, 399–409. [CrossRef]
51. Sheppard, S.R.; van den Bosch, C.C.K.; Croy, O.; Macias, A.; Barron, S. Urban forest governance and community engagement. In *Routledge Handbook of Urban Forestry*; Taylor & Francis: Abingdon, UK, 2017; pp. 205–221.
52. Barron, S.; Nitoslawski, S.; Wolf, K.L.; Woo, A.; Desautels, E.; Sheppard, S.R. Greening Blocks: A Conceptual Typology of Practical Design Interventions to Integrate Health and Climate Resilience Co-Benefits. *Int. J. Environ. Res. Public Health* **2019**, *16*, 4241. [CrossRef] [PubMed]
53. Kardan, O.; Gozdyra, P.; Misic, B.; Moola, F.; Palmer, L.J.; Paus, T.; Berman, M.G. Neighborhood greenspace and health in a large urban center. *Sci. Rep.* **2015**, *5*, 11610. [CrossRef] [PubMed]
54. Nowak, D.J.; Hirabayashi, S.; Bodine, A.; Greenfield, E. Tree and forest effects on air quality and human health in the United States. *Environ. Pollut.* **2014**, *193*, 119–129. [CrossRef] [PubMed]
55. Richardson, E.A.; Pearce, J.; Mitchell, R.; Kingham, S. Role of physical activity in the relationship between urban green space and health. *Public Health* **2013**, *127*, 318–324. [CrossRef] [PubMed]

© 2020 by the authors. Licensee MDPI, Basel, Switzerland. This article is an open access article distributed under the terms and conditions of the Creative Commons Attribution (CC BY) license (http://creativecommons.org/licenses/by/4.0/).

Article

Sensitivity Analysis of Machine Learning Models for the Mass Appraisal of Real Estate. Case Study of Residential Units in Nicosia, Cyprus

Thomas Dimopoulos [1,2,*] and Nikolaos Bakas [1]

1. School of Architecture, Land & Environmental Sciences, Department of Real Estate, Neapolis University Pafos, 2 Danais Avenue, 8042 Pafos, Cyprus; n.bakas@nup.ac.cy
2. Faculty of Engineering & Technology, Department of Civil Engineering & Geomatics, Cyprus University of Technology, P.O.Box 50329-3603, Limassol, Cyprus
* Correspondence: t.dimopoulos@nup.ac.cy or thomas.dimopoulos@cut.ac.cy

Received: 30 October 2019; Accepted: 15 December 2019; Published: 17 December 2019

Abstract: A recent study of property valuation literature indicated that the vast majority of researchers and academics in the field of real estate are focusing on Mass Appraisals rather than on the further development of the existing valuation methods. Researchers are using a variety of mathematical models used within the field of Machine Learning, which are applied to real estate valuations with high accuracy. On the other hand, it appears that professional valuers do not use these sophisticated models during daily practice, rather they operate using the traditional five methods. The Department of Lands and Surveys in Cyprus recently published the property values (General Valuation) for taxation purposes which were calculated by applying a hybrid model based on the Cost approach with the use of regression analysis in order to quantify the specific parameters of each property. In this paper, the authors propose a number of algorithms based on Artificial Intelligence and Machine Learning approaches that improve the accuracy of these results significantly. The aim of this work is to investigate the capabilities of such models and how they can be used for the mass appraisal of properties, to highlight the importance of sensitivity analysis in such models and also to increase the transparency so that automated valuation models (AVM) can be used for the day-to-day work of the valuer.

Keywords: general valuation; Cyprus; artificial intelligence; mass appraisals; real estate; algorithms; mathematical models; AVM; CAMA

---

1. Introduction

1.1. Background of the Study

Machine Intelligence imitates human perception, by utilizing mathematical models that compete against humans to deliver certain tasks such as the assessment and analysis of a studied system and predictions of out-of-sample observations. The accomplished tasks can be highly complex, based on mathematical models which simulate a physical, social, financial and so forth, system of study [1,2]. Machine learning algorithms belong to the wider thematic area of Artificial Intelligence, with applications in Healthcare [3], Automotive (self-driving cars) [4], Finance and Economics (predictions, assets management) [5], Military (drones capable of autonomous action) [6], Advertising (predict/quantify the behaviour of customers) [7], Image Recognition [8], and so forth. The idea that machines could exhibit intelligence is not a new concept, rather it stems from ancient times [9], for example, the robot Talos made by Hephaestus in Ancient Greek Mythology [10]. A bibliometric study of Artificial Intelligence Algorithms in Mass Appraisals Research [11] revealed that complex methods are increasingly considered in Real Estate predictions, in contrast to the well-established five methods for valuations.

At this point, it should be noted that remote sensing, aerial or oblique photos can be used in order to obtain this information automatically (this research paper though, only focuses on the mathematical modelling of price prediction. It should also be noted, however, that professional valuers hesitate to utilize such algorithms [12], and certain questions arise concerning the application of mathematical models for the improvement of the valuer's work. Questions such as, whether Machine Intelligence can replace Human Intelligence, if the mathematical models can replace the judgment of the individual valuer, and who would sign the outcome of an Automated Valuation Model are commonly debated topics. Yann LeCun quoted: "Our intelligence is what makes us human, Artificial intelligence is an extension of that quality. Many discussions have been had over recent years about whether there shall be a limit to restrict artificial intelligence and which level of artificial intelligence is optimal". Merriam-Webster [13] would add that AI can perform tasks that a human is unable to perform either at the same pace, quality, at the same cost or at all. The question arises whether artificial intelligence can replace the human valuer, taking into consideration computer-assisted mass appraisal (CAMA) and automated valuation models (AVM). It needs to be stated that, within the environment of appraisals, CAMA and AVM have been used since as early as the 1950s [14], and were further developed in the 1960s.

1.2. State of the Art

It appears that machine intelligence up until today can only successfully replace humans in the execution of specific tasks [15,16] which are often repetitive, dull and time-consuming [17]. Typical examples of Machine Intelligence for the case of Real Estate Valuations could be the collection of comparable evidence, automated exclusion of incorrect registrations in a database (anomaly detection) [18], calculation of the uncertainty of prediction in each particular chosen region through the computation of the local outliers and calibration of the prediction according to spatial parameters [19]. Contradictory, Artificial Intelligence cannot be considered to accurately understand specific property characteristics regarding the quality of construction, aesthetic characteristics, design, internal materials and appliances, the view to sea or nature, the deterioration of specific structural elements, local price peaks where comparable evidence is not available, property ownership issues (shares of ownership, rights of use etc.) and tax, legal or governmental special cases, because such models are complex and incomprehensible [20].

Artificial intelligence and machine learning methods have been widely utilized in Real-Estate, and a variety of studies have been performed. In Reference [21] the Hierarchical Linear Model is utilized in Mass appraisals of residential properties, to overcome the limitations of traditional econometric models such as Ordinary Least Squares. The absence of data of comparable properties is a major issue, while the consideration of micro- & macro- level characteristics of the properties should be considered [22]. In Reference [23], the spatial and temporal variation of properties is investigated, by a regression-cokriging method. However, to the best of our knowledge, no study exists on the interpretation of the black-box machine learning models, regarding Real Estate Mass Appraisals.

The purpose of this study is to investigate how the complex, machine learning models work, regarding Real Estate price predictions, and present the various models and the corresponding results. In Section 2, we explain the analyzed dataset as well as it variables, followed by the Machine Learning Methods utilized for the target task, as well as the generic algorithm to obtain the closed-form formula for the Higher Order Regression Model, via an automated, step-wise method. In Section 3, we present the sensitivity analysis results of the predictors, regarding Real Estate prices. In Section 3.3, the influence of the dataset volume is also investigated, by a parametric study, for a variety of partitions of the given dataset. In Section 3.4, we present the obtained formulas, utilizing five (5), and ten (10), and in Appendix A.1, for one hundred (100) nonlinear terms.

2. Comparable Evidence and Methods

2.1. Database, Pre-Processing, Methods and Performance Metrics

The studied database was obtained from the Department of Lands and Surveys (DLS). The data were used for the purposes of the Cyprus new general valuation [24] and refers to transactions between 2008 and 2014, out of which only transactions for apartments in Nicosia District were studied. Although it does not contain important socioeconomic variables [25], it is considered as vastly useful by professional valuers, as it contains comparable evidence about certain property types. Hence, the level of information available for the valuer could be greatly enhanced; however, the reliable exploitation of the contained information remains vague. A significant effort was spent in order to prepare the database in a predictors-output format. At this point, the authors highlight that the data would be significantly enhanced if remote sensing was integrated in order to enrich the database provided that was completed by on-site or drive-by observations.

In particular, 4261 observations of apartment/office sales in Nicosia existed, nevertheless from column Unit_desc, only values "APARTMENT" & "2-FLOOR APPRTMENT" were kept, resulting in 3786 remaining observations. Furthermore, only Municipalities that are regulated by the Nicosia Local Town Plan were selected, those Quarters with less than 20 observations were deleted and, finally, 3561 sales data were used for the analysis and predictions. In order to enhance the prediction accuracy of the models, Urban Planning data were added for each Planning Zone, and in particular, the maximum building density, the number of stories, height and coverage of the allowed building, the minimum sq.m. per resident and the expected sq.m. per resident. Due to multicollinearity among urban planning variables, only the maximum building density was finally kept. The transaction dates were converted to reflect the date 30 September 2018, as floating numbers constituting a continuous variable, and the prices were adjusted to 1 January 2013 utilizing the Central Bank of Cyprus Index. This index is using property data gathered from valuations submitted to the contracted banks since 2006. The relevant information is provided from independent property surveyors that evaluate properties mainly for mortgage purposes such as housing loans, mortgage refinancing and mortgage collateral.

The utilized variables were as follows, with their abbreviations in parentheses, for each Unit (Appartment)

- Unit Enclosed extent, which is the Internal Area in m^2 (IntArea).
- The Unit covered extent, which is the Area of covered verandahs in m^2 (CovVer).
- The Unit uncovered extent, which is the Area of uncovered verandahs in m^2 (UnCovVer).
- Parcel extent, that is the Area of parcel (or plot) in m^2 (ParcExt).
- The Built Years, calculated as the difference among the date the transaction happened and the date the building was constructed, in years (BuiltYrs).
- The Unit condition code (Cond), that denotes the condition of the building, and takes values from 1 (best condition) to 4 (worst condition).
- The Unit's view code (View), which denotes the view of the unit, with values from 1 (best view) to 4 (worst view).
- The Unit's class code (Class), denoting the class of the building. It takes Values from 1 (best class) to 4 (worst class).
- Density (Dens), as the maximum allowed density (built m^2, over plots m^2) of the specific district.

The dependent variable was the apartment's price as accepted by the Cyprus Department of Lands and Surveys. This price was adjusted by utilizing the Central Bank of Cyprus Index and the dates were transferred to 30 September 2018. The abbreviation for the dependent variable is (Adj. Accepted Price).

2.2. Error Metrics

Machine learning methods exhibit diverse performance on a studied dataset, with respect to the error metrics each time utilized. The Coefficient Of Dispersion (COD) was used (Equation (1)) as defined by Appraisal Ratio Studies [26], as a common metric utilized in Real Estate Appraisals. It is based on the Predicted Values (PV), the Dependent Variable (DV), and the number of observations N. COD is defined by $COD = 100 \frac{\frac{1}{N}\Sigma(|\frac{PV}{DV}| - \frac{1}{N}\Sigma\frac{PV}{DV})}{\frac{1}{N}\Sigma\frac{PV}{DV}}$. Furthermore, the utilized error metrics were the Root Mean Squared Error $RMSE = \sqrt{\frac{\Sigma(PV-DV)^2}{N}}$, the Mean Absolute Error $MAE = \frac{\Sigma|PV-DV|}{N}$, the Mean Absolute Percentage Error $MAPE = \frac{1}{N}\Sigma\frac{|PV-DV|}{DV}$, the Maximum Absolute Percentage Error (MAXAPE), as well as the Pearson Correlation Coefficient ρ, the slope of the Predicted versus Actual values α, such that $PV = \alpha * DV + \beta$, and the $SR = \frac{1}{N}\Sigma\frac{PV}{DV}$.

2.3. Anomaly Detection

Although the observations in the studied database regard official registration in the DLS, some extremely unreasonable records occur. For example, property in Nicosia Municipality, Ag. Andreas Quarter, built in 1965, with 66 sq.m covered area, and a price of 3.524€, Latsia/Ag. Georgios (1977), 68 sq.m, with a price of 17.781 €, Nicosia/Ag. Omologites (1982), 44 sq.m, 15.724 €, Nicosia/Ag. Antonios (1973), 35 sq.m, 22.562 €, and. Strovolos/Chryseleousa (1986), 76 sq.m, 17.283 €. Accordingly, an iterative procedure was implemented in order to identify the outliers and eliminate at each step the observation which violates a specified threshold. The corresponding results were highly enhanced, as even for the Linear Regression (LR) (Figure 1) the R squared was increased from 0.611 to 0.744, while the shape of the scattered observations is closer to a straight line after the removal of the outliers.

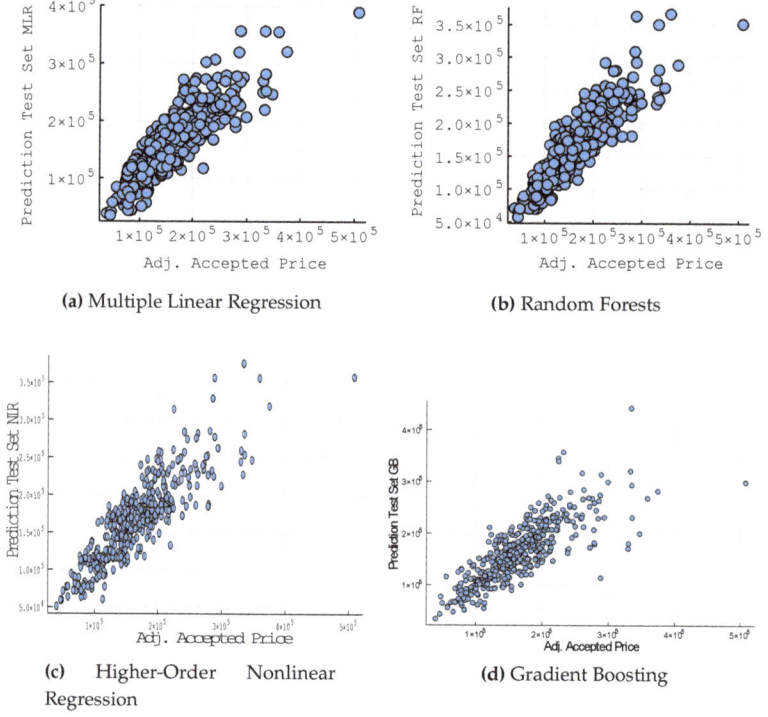

Figure 1. Accepted Price vs. Simulated for the test-set data.

Algorithm 1 was selected in order to exclude observations with high prediction errors, as they represent apartments which were under- or over- priced by the DLS, for some particular reason. The algorithm was selected amongst others because it presented better results in terms of percentage errors that are more easily understood by property professionals.

Algorithm 1: Anomaly Detection

Data: Predictors($IntArea, CovVer, UnCovVer, ParcExt, BuiltYrs, Cond, View, Class, Dens$), and Responce ($Adj.AcceptedPrice$) for the entire Dataset.
Result: New, Decreased Dataset
1 Do Linear Regression
2 Compute $MAXAPE$
3 **while** $MAXAPE \geq 50\%$ **do**
4 $\quad$ Linear Regression;
5 $\quad$ Compute $MAXAPE$;
6 $\quad$ Find index i of observation with $APE_i = MAXAPE$;
7 $\quad$ Delete observation with index i;
8 **end**
9 **return** Decreased Dataset

2.4. Machine Learning Methods

In order to evaluate more complex models, apart from Multiple Linear Regression (MLR), a Higher Order, Nonlinear Regression (NLR) was implemented. In particular, all combination of the variables were created, up to third order

$$x_i * x_j * x_k,$$

with $i, j, k \in [1, 9]$ for all the nine independent variables. Afterwards, a forward step-wise algorithm was implemented, in order to sequentially add to the model the combined variable with $x_i * x_j * x_k$, which corresponds to the model with the lowest APE. Algorithm 2 represent the applied procedure.

Algorithm 2: Step-wise, Higher Order Regression

Data: Predictors($IntArea, CovVer, UnCovVer, ParcExt, BuiltYrs, Cond, View, Class, Dens$), Responce ($Adj.AcceptedPrice$) for the Decreased Dataset, and desired number of features n_f
Result: NLR Model
1 Create Nonlinear Features $x_i * x_j * x_k$
2 Compute all APE with Uni-variate Regression
3 Store i, j, k combination corresponding to the minimum APE
4 **for** $i_1 = 2 : n_f$ **do**
5 $\quad$ **for** $i_2 = 1 :$ all non-Stored features **do**
6 $\quad\quad$ Add i_2 to the Model
7 $\quad\quad$ Compute all APE with Multi-variate Regression
8 $\quad$ **end**
9 $\quad$ Store i, j, k combination corresponding to the i_2, with minimum APE
10 **end**
11 **return** all combinations of i, j, k for the NLR Model

Furthermore, we utilized Random Forests (RF) [27] as implemented in Reference [28], and Gradient Boosting (GB) [29]. All analyses were run on Juia [30] programming language by utilizing the mentioned packages, as well as code written by the authors, as described in Algorithms 1 and 2.

3. Results

3.1. Regression Analysis

The regression results are presented in Table 1 for the four methods studied and the corresponding error metrics.

Table 1. Regression Results for the four methods studied, and error metrics.

Methods	ρ	MAE	RMSE	MAPE	MAXAPE	SR	α	COD
				Train Set				
Random Forests	0.914	17931.100	28854.237	0.111	1.307	1.031	0.739	10.778
Gradient Boosting	0.992	2630.784	8923.668	0.016	0.441	1.002	0.983	1.753
Linear Regression	0.863	24546.300	34745.422	0.151	0.550	1.027	0.746	14.703
Non-Linear Regression	0.880	23520.570	32700.793	0.146	1.100	1.032	0.775	14.197
				Test Set				
Random Forests	0.877	20817.165	27950.722	0.134	0.802	1.040	0.753	12.950
Gradient Boosting	0.803	24485.519	35946.437	0.151	1.092	1.009	0.776	15.017
Linear Regression	0.858	22977.825	30047.707	0.146	0.506	1.025	0.789	14.279
Non-Linear Regression	0.862	22525.779	29500.974	0.144	0.552	1.032	0.761	13.984

3.2. Sensitivity Analysis

A modified version of the Profile method [31,32] is utilized, in order to investigate the contribution of each independent variable to the dependent variable. In particular, each input variable varies within its given (raw) range while all the other input variables are kept constant in a certain value. This constant takes three discrete values: 25% Percentile, Median, 75% Percentile. Through Sensitivity Analysis, the comparison of the black-box models can be illuminated, as we compare the effect of a predictor (i.e., Unite Enclosed Extent in Figure 2) to the studied variable (Adj. Accepted Price), indicating a decreasing pattern for IntArea higher than 180 m^2, which cannot be identified with the Linear Model. Accordingly, in Figure 3, the Adj. Accepted Price is being decreased with respect to the built years; however, the Machine Learning models concur that this effect is weakens for built years more than 30. However, although all models exhibit similar patterns, different sensitivity curves are obtained for each model. This effect indicates the complexity of such models, which should be utilized critically, or as ensembles [33]. The complete presentation of the Sensitivity Analysis Figures for all predictors is presented in Appendix A.

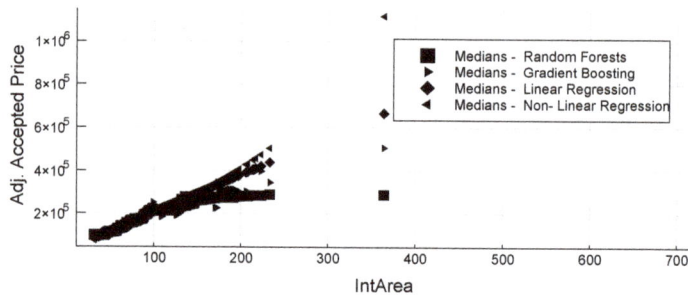

Figure 2. Sensitivity Analysis for Unit Enclosed Extent

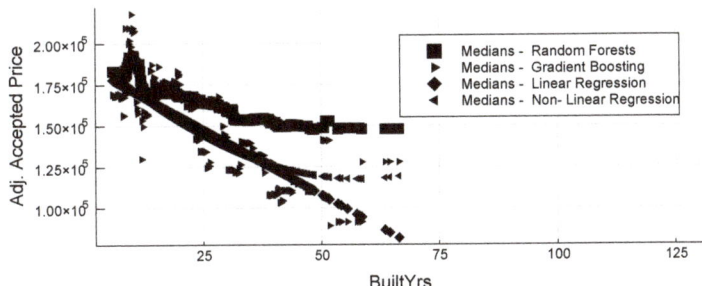

Figure 3. Sensitivity Analysis for Unit Built Years.

3.3. How Much Data Is Big Enough?

A common problem in simulation with Machine Learning Methods is the amount of data. In order to investigate the importance of the data volume to the accuracy of prediction, we utilized random portions of the dataset and each time we fitted a Random Forests model to the partition of the data. In Figure 4 we present the corresponding Mean Absolute Percentage Errors concerning the number of observations. Afterwards, we fit a logarithmic curve:

$$y = \alpha \log(x) + \beta, \tag{1}$$

to the obtained results, and extended the curve up to 5000 observations from the results we see that the number of data is an important factor influencing the prediction accuracy, with a clear decreasing pattern.

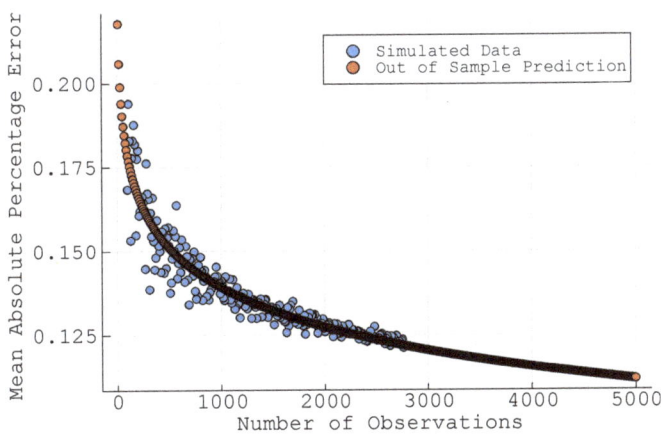

Figure 4. Number of data importance

3.4. Prediction Formula

With nonlinear Regression, we obtain the following Equations for the prediction:

(A) With five terms (MAE=23993 €)

$$\begin{aligned}
Adj.AcceptedPrice = &\ 2.13785E + 03 * IntArea \\
&- 2.44629E + 01 * BuiltYrs * IntArea \\
&+ 4.67313E - 01 * BuiltYrs * CovVer * IntArea \\
&+ 3.52720E + 02 * UnCovVer \\
&+ 2.43798E + 02 * Dens * View * CovVer + 1.09116E + 04
\end{aligned} \quad (2)$$

(B) With ten terms (MAE=23748 €, also used in sensitivity)

$$\begin{aligned}
Adj.AcceptedPrice = &\ 2.87808E + 03 * IntArea \\
&- 3.52523E + 01 * BuiltYrs * IntArea \\
&+ 1.35281E - 02 * BuiltYrs * CovVer * IntArea \\
&+ 3.30431E + 02 * UnCovVer + 5.16573E \\
&+ 02 * Dens * View * CovVer + 3.01148E \\
&- 01 * BuiltYrs * BuiltYrs * IntArea \\
&+ 2.32119E - 02 * IntArea * IntArea * IntArea \\
&- 6.87503E + 00 * IntArea * IntArea \\
&- 1.57789E + 00 * View * Cond * ParcExt \\
&+ 3.23269E - 04 * ParcExt * ParcExt * Dens - 7.24500E + 03
\end{aligned} \quad (3)$$

4. Discussion

Sensitivity analysis for features' importance to the dependent variable (Adj. accepted Price), demonstrated similar patterns, for all the four methods used. However, Certain differences were also depicted, which highlights the need for such analyses on the trained machine learning models. The accurate modelling of a studied system is challenging, and its predictive value is controversial [12,34], while the hopeful prospects that computers and refined models, will accomplish high prediction accuracy, were repeatedly defeated [1]. The utilization of a more accurate model instead of empirical rules exhibited enhanced prediction accuracy in property valuations. However, mathematical models without error estimation could jeopardize valuations hence we recommend that one obtains an initial estimation +/− a prediction error, as well as comprehensively investigating the errors' extrema and distributions. Machine learning algorithms can be used to validate professional valuations and not to replace human judgment, in order to avoid the impact of the highly improbable [35].

The outermost important factors that the authors recommend to be examined are Time, Money, Quality, Accuracy, Bureaucracy, Responsibility, Regulations, Licenses, Initial cost, Neutrality and available data. Every single property valuation is a unique project and has a clear starting and ending date. Manual valuations are usually resourced intensive for both time and money and often deliver results in crucial revaluations later or sometimes never (Quevara [36]). In a project, there is always a trade-off between Time, Money and Quality. Increasing one of the factors almost automatically decreases the remaining two. For example, a valuer who tries to complete more valuations within a given period, either must decrease the quality of each valuation to be faster per valuation or must hire more staff to deliver more valuations. AI does not have any of these constraints. It can work 24/7 and with the correct data, can produce a theoretically infinite amount of valuations. Practically, the amount is limited to the available data as well as the input of this data by a human source.

In the above paragraph, data has been mentioned as an important component. CAMA and AVM can only exhibit high computational efficiency if the database contains adequate data. Theoretically, one could state that if no data is available, AI could not be used. On the other hand, without precise data, any human-based valuation would not be very precise either. It takes years of studying and obtaining practical experience as well as local market knowledge for a valuer to be able to deliver accurate valuations and appraisals. This process of learning is time-consuming and rather expensive. AI can do so within a short period of time and can improve its performance based on past observations. Due to that, human valuers are expensive. AI can offer a much less expensive rate for any valuation since cost such as travel time and travel expenses to the property can be saved. However, AI has a higher initial cost as it is expensive to set up a model. The maintenance of the database and feeding the AI model with more data are usually the highest running expenses. Any invention that may replace workers with machines in a particular field can have a positive effect on society by "reducing the price of goods, increasing real income" [37]. Research conducted in this context suggests that the methods, currently used extensively, have inherent errors regarding how they derive their value estimates [38]. Many scientists stated that feelings and sympathy are what make us humans. These are unarguably great assets of every human; however, in valuations, they can create inaccuracies due to the loss of neutrality. Humans can only control their doings up to a certain level. AI does not lose neutrality and hence accuracy, due to sympathy, therefore, in this aspect it can create more accurate valuations.

Carrying out an official valuation requires, in almost every country, a license. These licenses are often provided by human-based associations. Often political reasons block any technological process as some humans fear losing their job to AI. This political lobbying reduces progress considerably and by doing so the human valuer is heavily favoured. Human valuers often argue about the responsibility and legal pursuit of AI. A valuation carried out by a human valuer can always be challenged and one can sue the person who completed the valuation but the questions to be answered are—who do you sue when a CAMA valuation is in question, and who signs a CAMA valuation. The above two questions can unfortunately not be answered easily. Looking for the responsible party of a CAMA valuation is a tricky process, which is one of the major drawbacks of AI. However, if we feed the AI model with enough data and constantly maintain and update the database, the possible margin of error shall be small enough to be negligible, and costly legal processes could be avoided or minimized. Besides that, we must understand in which situations we value properties and if all valuations need to be legally appropriate in terms of responsibility and suitability. Nowadays, countless valuations are done daily; mostly valuations for courts or banks giving out mortgages or attempting to repossess distressed/mortgaged assets, but there are so many more valuations conducted for many other reasons.

All the explanations described in the above paragraph could be ideal situations for the use of AI, in order to provide cheaper and faster valuations. Having this kind of valuation completed by AI models would, of course, reduce the total number of valuations completed by human valuers. However, it has to be stated that the effect of artificial intelligence on the level of human employment will be dramatic reduced [39]. This, however, does not necessarily mean that any human valuer should lose their job. It could mean the opposite. Human valuers could focus more on each valuation, automatically increasing the quality of every valuation completed by a human valuer. Special reference must be made to complex valuations where a valuer needs a lot of time to fully understand and adjust the influencing factors. By giving human valuers more time to focus on these complex valuations and valuations for bank lending or repossessing purposes, increases the quality significantly. The improvement of quality will automatically lead to a higher achieved price per valuation which could, in the end, create higher profits for any valuer.

Remote Sensing Integration in Mass Appraisals

Remote sensing is another important tool that can be used in Mass Appraisals and data collection. In remote sensing, information about a given category of property is acquired without necessarily visiting the property [40]. According to Nayak and Zlatanova [41], remote sensing experts establish GIS systems that are often utilized. Remote sensing makes it possible to determine the attributes of a property such as its location, lot size, and type of structures that have been erected on the land. This is especially helpful because some property may be located in areas where access is restricted, as mentioned by Xiao-sheng, Zhe and Ting-li [42]. Remote sensing makes the identification of property easier because in the remote sensing developed maps, property lines can be drawn that show the exact location of the property [43]. Remote sensing can also be used to provide measures for a number of dependent variables, which are linked to human activity, especially with regards to the environmental impacts of various social, economic, as well as, demographic processes. For instance, remote sensing observations of land cover may depict the footprints of agricultural intensification, the expansion of urban areas, as well as road development and many other factors that are affecting the value of properties. These may also entail observations of vegetation density that may be linked to the impacts of fertilization, irrigation, coupled with other agricultural practices. Other areas may cover observations of new buildings constructions that are related to mass appraisals. Therefore, models that combine remote observations with ground-based social data may be very important in understanding their market value.

5. Conclusions

Machine learning models are highly non-transparent and it is difficult to completely understand what affects the value of a particular property the most. We defeat this issue by detailed sensitivity analysis for each predictor, by utilizing and comparing four machine learning models. Further studies in this sector need to be carried out in order to improve the overall transparency of any model used. However, Machine learning models are characterized by a consistent error across all the given observations, which follows a known statistical distribution, while valuations completed by human valuers might contain different types and magnitude of biases. The models would be even more precise if the database was enriched with more data that are related to the characteristics of the property. The easiest and cheapest way to get these data today is through satellite imagery. Data such as elevation, building height, age, construction type and distance from value influence centers such as schools, hospitals, public transportation and so forth, or even pollution or air quality in the area under study can be collected from satellites. Lastly, with machine learning techniques, important constraints have been identified such as the transparency of models and the repeatability of the results [14]. Especially in Cyprus, larger-scale tests on still needed to be completed repeatedly. Finally, machines have already taken over a lot of jobs that were previously carried out by humans and every time we got to a point where the chance that humans could lose jobs, more jobs were created thereby increasing prosperity and the quality of life for humans. Machines assist us and improve our lives. Coming back to the starting quote, machines, and especially AI as described above, are capable of increasing our quality of intelligence as humans.

Author Contributions: T.D. conceived of the presented idea and wrote the main part of the manuscript. N.B. performed the computations and wrote the relevant part. T.D. interpreted the numerical results.

Funding: This research received no external funding.

Acknowledgments: Special thanks to Varnavas Pashoulis from the Department of Lands and Surveys, who helped us to gain access to the dataset used in this research paper. Also to D.G. Hadjimitsis for his continuous support.

Conflicts of Interest: The authors declare no conflict of interest.

Appendix A

Appendix A.1. Prediction Formula with 100 terms (MAE = 19694 €)

$Adj.AcceptedPrice = 3.31113E + 03 * IntArea - 4.68398E + 01 * BuiltYrs * IntArea - 5.65708E - 01 * BuiltYrs * CovVer * IntArea + 1.86782E + 03 * UnCovVer - 2.78991E + 02 * Dens * View * CovVer + 4.88587E - 01 * BuiltYrs * BuiltYrs * IntArea + 6.60039E - 02 * IntArea * IntArea * IntArea - 1.70363E + 01 * IntArea * IntArea + 3.24869E + 00 * View * Cond * ParcExt + 1.70068E - 03 * ParcExt * ParcExt * Dens + 1.30015E + 01 * Cond * BuiltYrs * IntArea - 5.17967E - 07 * ParcExt * ParcExt * ParcExt - 1.00466E + 01 * Dens * BuiltYrs * IntArea - 5.65825E - 02 * View * BuiltYrs * ParcExt - 8.15277E + 00 * Class * ParcExt + 4.49178E - 02 * BuiltYrs * UnCovVer * UnCovVer + 6.90546E +$

$01 * CovVer * IntArea - 1.04803E + 02 * CovVer * CovVer - 1.55104E - 03 * ParcExt * UnCovVer * UnCovVer - 1.27722E + 01 * Dens * BuiltYrs * BuiltYrs + 3.00378E + 03 * BuiltYrs - 1.23882E + 01 * Cond * CovVer * BuiltYrs - 5.87984E + 01 * BuiltYrs * BuiltYrs + 5.66044E - 01 * BuiltYrs * BuiltYrs * BuiltYrs - 1.16139E + 02 * Class * Class * UnCovVer - 1.00749E + 03 * Class - 1.54169E + 02 * Class * Cond * BuiltYrs - 5.23747E - 02 * CovVer * CovVer * CovVer - 1.69265E - 01 * CovVer * IntArea * IntArea - 3.02085E + 03 * CovVer + 1.27557E + 01 * Dens * ParcExt * Dens + 8.11517E + 04 * Dens + 2.19071E + 00 * Cond * UnCovVer * UnCovVer - 2.34493E + 02 * Class * UnCovVer - 7.94913E - 01 * BuiltYrs * CovVer * CovVer + 3.26919E - 01 * Cond * ParcExt * CovVer - 7.49578E - 06 * ParcExt * ParcExt * IntArea + 1.38552E + 02 * Dens * View * IntArea - 2.01899E + 03 * View * View - 4.53539E - 01 * Class * ParcExt * CovVer - 2.15116E + 00 * Cond * IntArea * IntArea - 3.77756E - 01 * Cond * BuiltYrs * ParcExt + 9.21765E - 01 * Dens * BuiltYrs * ParcExt + 1.84117E - 03 * ParcExt * IntArea * IntArea - 3.30262E - 01 * ParcExt * IntArea + 3.04830E + 00 * Cond * Cond * ParcExt - 1.46508E + 01 * Dens * Cond * ParcExt - 3.86245E - 02 * UnCovVer * IntArea * IntArea + 5.24053E - 01 * UnCovVer * CovVer * CovVer - 2.75663E + 01 * Cond * UnCovVer * CovVer + 7.11439E + 01 * View * BuiltYrs * UnCovVer + 1.02919E - 01 * UnCovVer * UnCovVer * IntArea - 3.28336E + 00 * Class * UnCovVer * UnCovVer - 7.87081E + 02 * Dens * Cond * CovVer - 2.26100E + 03 * View * View * Cond + 5.63615E - 01 * UnCovVer * CovVer * BuiltYrs + 4.64562E + 01 * View * Cond * Cond + 2.81664E + 01 * CovVer * CovVer * View - 1.93202E - 01 * View * CovVer * ParcExt - 4.70719E + 00 * Dens * BuiltYrs * UnCovVer + 6.68377E + 03 * Dens * View * Class + 3.48819E + 01 * View * UnCovVer * CovVer + 5.50960E + 01 * Dens * UnCovVer * CovVer + 6.22286E + 01 * Dens * UnCovVer + 8.19344E + 02 * Dens * Class * UnCovVer - 4.44616E + 04 * Dens * Dens - 1.24587E + 01 * Dens * Class * ParcExt + 5.88687E + 00 * Class * Class * ParcExt - 1.66661E - 01 * CovVer * CovVer * IntArea + 6.65160E + 03 * Dens * Dens * Cond + 3.01051E + 01 * Cond * CovVer * CovVer + 1.08769E + 02 * BuiltYrs * CovVer - 2.07236E + 01 * View * BuiltYrs * CovVer - 5.19018E + 00 * Cond * BuiltYrs * UnCovVer - 1.55653E - 01 * UnCovVer * IntArea * BuiltYrs + 1.09671E + 00 * Dens * ParcExt * CovVer - 1.29912E + 02 * View * BuiltYrs * Dens - 7.65347E + 00 * BuiltYrs * UnCovVer * Class + 9.07541E + 01 * Dens * Cond * IntArea - 4.31216E - 04 * ParcExt * UnCovVer * CovVer - 1.37379E + 01 * View * UnCovVer * UnCovVer + 5.77048E - 03 * ParcExt * ParcExt - 5.29492E - 03 * BuiltYrs * BuiltYrs * ParcExt - 1.10686E + 03 * Class * Class * View + 6.79802E - 01 * Dens * UnCovVer * IntArea - 2.49660E + 00 * Dens * View * ParcExt + 1.30759E + 03 * Class * CovVer - 1.71103E + 02 * Class * CovVer * Class + 9.43679E + 01 * Cond * CovVer * Cond + 2.89773E + 01 * Dens * Dens * BuiltYrs - 1.90894E + 02 * UnCovVer * CovVer - 1.54163E + 03 * Dens * View * UnCovVer - 2.77202E + 03 * Dens * Dens * Dens + 5.71547E + 00 * Dens * BuiltYrs * CovVer + 2.71628E + 02 * View * UnCovVer * View + 2.80269E + 01 * UnCovVer * UnCovVer - 7.70039E + 01 * BuiltYrs * UnCovVer - 2.44312E + 00 * Dens * UnCovVer * UnCovVer + 2.38252E + 01 * Class * UnCovVer * CovVer - 4.06621E - 05 * ParcExt * ParcExt * CovVer - 7.10543E + 04$

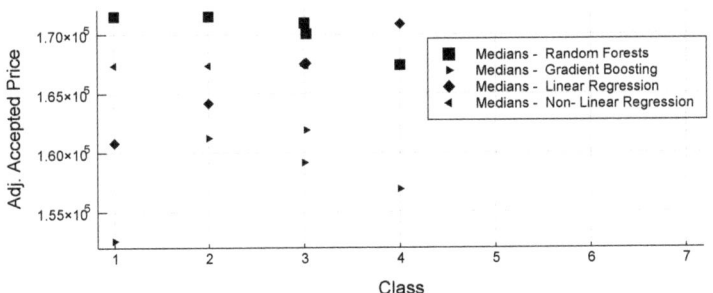

Figure A1. Sensitivity Analysis for Unit Class.

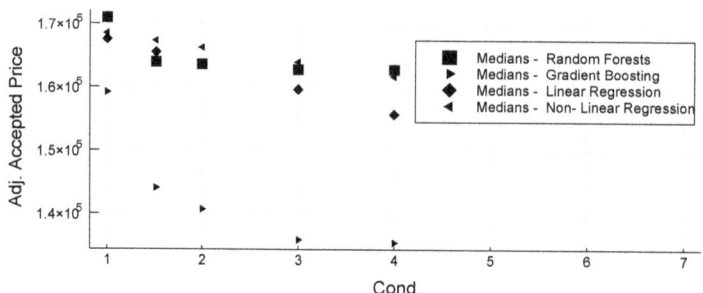

Figure A2. Sensitivity Analysis for Unit Condition Code.

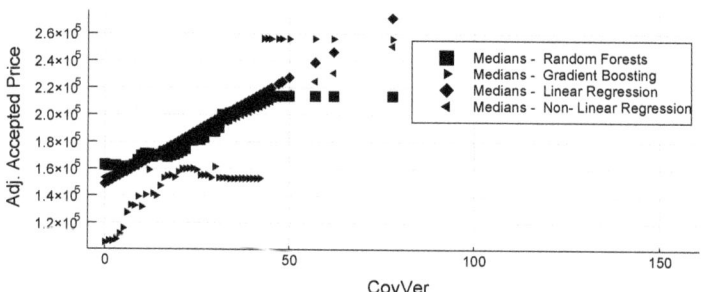

Figure A3. Sensitivity Analysis for Unit Covered Extent.

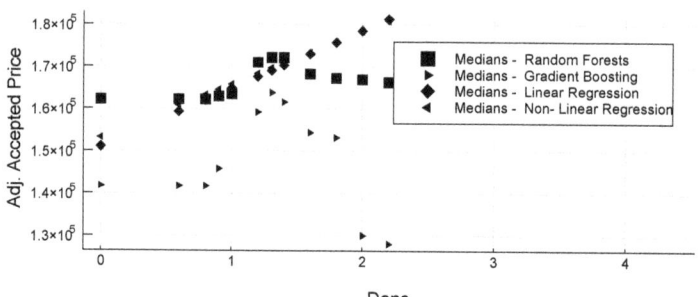

Figure A4. Sensitivity Analysis for Density.

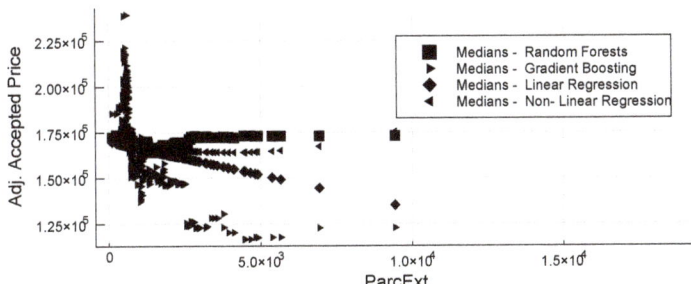

Figure A5. Sensitivity Analysis for Parcel Extent.

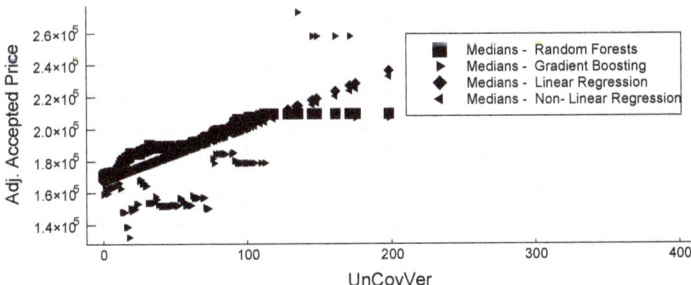

Figure A6. Sensitivity Analysis for Uncovered Extent.

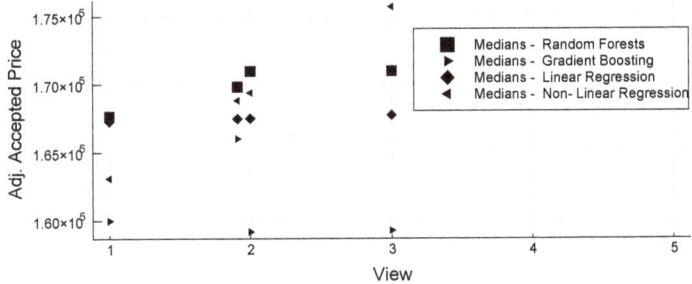

Figure A7. Sensitivity Analysis for View.

References

1. Makridakis, S.; Bakas, N. Forecasting and uncertainty: A survey. *Risk Decis. Anal.* **2016**, *6*, 37–64. doi:10.3233/RDA-150114. [CrossRef]
2. Bakas, N.P. Numerical Solution for the Extrapolation Problem of Analytic Functions. *Res.* **2019**, *2019*, 1–10. doi:10.34133/2019/3903187. [CrossRef] [PubMed]
3. Jiang, F.; Jiang, Y.; Zhi, H.; Dong, Y.; Li, H.; Ma, S.; Wang, Y.; Dong, Q.; Shen, H.; Wang, Y. Artificial intelligence in healthcare: past, present and future. *Stroke Vasc. Neurol.* **2017**, *2*, 230–243. [CrossRef] [PubMed]

4. Pozna, C.; Antonya, C. Issues about autonomous cars. In Proceedings of the 2016 IEEE 11th International Symposium on Applied Computational Intelligence and Informatics (SACI), Timisoara, Romania, 12–14 May 2016; pp. 13–18.
5. Gupta, S.; Sharma, A.; Abubakar, A.; others. Artificial intelligence–driven asset optimizer. In Proceedings of the SPE Annual Technical Conference and Exhibition, Dallas, TX, USA, 24–26 September 2018.
6. De Swarte, T.; Boufous, O.; Escalle, P. Artificial intelligence, ethics and human values: The cases of military drones and companion robots. *Artif. Life Robot.* **2019**, *24*, 291–296. [CrossRef]
7. Olson, C.; Levy, J. Transforming marketing with artificial intelligence. *Appl. Mark. Anal.* **2018**, *3*, 291–297.
8. Zadeh, L.A.; Tadayon, S.; Tadayon, B. System and Method for Extremely Efficient Image and Pattern Recognition and Artificial Intelligence Platform. U.S. Patent App. 15/919,170, 3 December 2018.
9. Cave, S.; Dihal, K. Ancient dreams of intelligent machines: 3,000 years of robots. *Nature* **2018**, *559*, 473–475. doi:10.1038/d41586-018-05773-y. [CrossRef]
10. MAYOR, A. *Gods and Robotss: Myths, Machines, and Ancient Dreams of Technology*; Princeton University Press: Princeton, NJ, USA, 2018. doi:10.2307/j.ctvc779xn. [CrossRef]
11. Dimopoulos, T.; Bakas, N. An artificial intelligence algorithm analyzing 30 years of research in mass appraisals. *Reland Int. J. Real Estate & Land Plan.* **2019**, *2*, 10–27.
12. Dimopoulos, T.; Tyralis, H.; Bakas, N.P.; Hadjimitsis, D. Accuracy measurement of Random Forests and Linear Regression for mass appraisal models that estimate the prices of residential apartments in Nicosia, Cyprus. *Adv. Geosci.* **2018**, *45*, 377–382. doi:10.5194/adgeo-45-377-2018. [CrossRef]
13. *Merriam Webster*; Merriam Webster: Miami, FL, USA, 2016.
14. Worzala, E.; Lenk, M.; Silva, A. An exploration of neural networks and its application to real estate valuation. *J. Real Estate Res.* **1995**, *10*, 185–201.
15. Brynjolfsson, E.; Mitchell, T.; Rock, D. What Can Machines Learn, and What Does It Mean for Occupations and the Economy? *Aea Pap. Proc.* **2018**, *108*, 43–47. doi:10.1257/pandp.20181019. [CrossRef]
16. Pagano, D. Machine Learning Will Replace Tasks, Not Jobs, Say MIT Researchers. *MIT News*, 26 June 2018.
17. Bryson, J.J. Robots should be slaves, 2010.
18. Agrawal, S.; Agrawal, J. Survey on anomaly detection using data mining techniques. *Procedia Comput. Sci.* **2015**, *60*, 708–713. [CrossRef]
19. Dimopoulos, T.; Moulas, A. A proposal of a mass appraisal system in Greece with CAMA system: Evaluating GWR and MRA techniques in Thessaloniki Municipality. *Open Geosci.* **2016**, *8*, 675–693. [CrossRef]
20. Chan, A.P.; Abidoye, R.B. Advanced property valuation techniques and valuation accuracy: Deciphering the artificial neural network technique. *Reland Int. J. Real Estate & Land Plan.* **2019**, *2*, 1–9.
21. Arribas, I.; García, F.; Guijarro, F.; Oliver, J.; Tamošiūnienė, R. Mass appraisal of residential real estate using multilevel modelling. *Int. J. Strateg. Prop. Manag.* **2016**, *20*, 77–87. [CrossRef]
22. Ciuna, M.; Milazzo, L.; Salvo, F. A mass appraisal model based on market segment parameters. *Building* **2017**, *7*, 34. [CrossRef]
23. Chica-Olmo, J.; Cano-Guervos, R.; Chica-Rivas, M. Estimation of housing price variations using spatio-temporal data. *Sustainability* **2019**, *11*, 1551. [CrossRef]
24. Cyprus New General Valuation 2013. 2013. Available online: https://portal.dls.moi.gov.cy/en-us/Pages/\New-General-Valuation-as-at-1-1-2013.aspx (accessed on 1 June 2019).
25. Lelo, K.; Tomassi, S.M.F. Urban inequalities in Italy: A comparison between Rome, Milan and Naples. *Entrep. Sustain. Issues* **2018**, *6*, 939–957. [CrossRef]
26. NCSS Statistical Software. *Appraisal Ratio Studies*; NCSS: Kaysville, UT, USA, 2015.
27. Breiman, L. Random forests. *Mach. Learn.* **2001**, *45*, 5–32. [CrossRef]
28. Sadeghi, B. DecisionTree.jl. 2013. Available online: https://github.com/bensadeghi/DecisionTree.jl (accessed on 1 June 2019).
29. Xu, B.; Chen, T. XGBoost.jl. 2014. Available online: https://arxiv.org/abs/1603.02754 (accessed on 1 June 2019).
30. Bezanson, J.; Edelman, A.; Karpinski, S.; Shah, V.B. Julia: A fresh approach to numerical computing. *Siam Rev.* **2017**, *59*, 65–98. [CrossRef]
31. Gevrey, M.; Dimopoulos, I.; Lek, S. Review and comparison of methods to study the contribution of variables in artificial neural network models. *Ecol. Model.* **2003**. doi:10.1016/S0304-3800(02)00257-0. [CrossRef]

32. Olden, J.D.; Jackson, D.A. Illuminating the "black box": A randomization approach for understanding variable contributions in artificial neural networks. *Ecol. Model.* **2002**. doi:10.1016/S0304-3800(02)00064-9. [CrossRef]
33. Sagi, O.; Rokach, L. Ensemble learning: A survey. *Wiley Interdiscip. Rev. Data Min. Knowl. Discov.* **2018**, *8*, e1249. [CrossRef]
34. Oreskes, N.; Shrader-Frechette, K.; Belitz, K. Verification, validation, and confirmation of numerical models in the earth sciences. *Science* **1994**. doi:10.1126/science.263.5147.641. [CrossRef] [PubMed]
35. Lybeck, E. The black swan: The impact of the highly improbable. *arXiv* **2017**, arXiv:1011.1669v3.
36. Bird, R.; Slack, E.; Guevara, M. Real Property Taxation in the Philippines. In *International Handbook of Land and Property Taxation*; Edward Elgar Publishing: Cheltenham, UK, 2013. doi:10.4337/9781845421434.00018. [CrossRef]
37. Autor, D.H. Why Are There Still So Many Jobs? The History and Future of Workplace Automation. *J. Econ. Perspect.* **2015**. doi:10.1257/jep.29.3.3. [CrossRef]
38. Ismail, S.; Buyong, T. *Residential Property Valuation Using Geographic Information System*; Bulgarian Geoinformation Company: Ovcha Kupel, Bulgaria, 1998.
39. Frey, C.B.; Osborne, M.A. The future of employment: How susceptible are jobs to computerisation? *Technol. Forecast. Soc. Chang.* **2017**. doi:10.1016/j.techfore.2016.08.019. [CrossRef]
40. Dimopoulos, T.; Labropoulos, T.; Hadjimitsis, D.G. Comparative analysis of property taxation policies within Greece and Cyprus evaluating the use of GIS, CAMA, and remote sensing techniques. *Proc. SPIE Int. Soc. Opt. Eng.* **2014**, *9229*, 92290O.
41. Nayak, S.; Zlatanova, S. *Remote Sensing and GIS Technologies for Monitoring and Prediction of Disasters*; Springer: Berlin, Germnay, 2008.
42. LIU, X.s.; Zhe, D.; WANG, T.l. Real estate appraisal system based on GIS and BP neural network. *Trans. Nonferrous Met. Soc. China* **2011**, *21*, s626–s630. [CrossRef]
43. Lindgren, D. *Land Use Planning and Remote Sensing*; Taylor & Francis: New York, NY, USA, 1984; Volume 2.

Sample Availability: The dataset was provided from the Department of Lands and Surveys.

© 2019 by the authors. Licensee MDPI, Basel, Switzerland. This article is an open access article distributed under the terms and conditions of the Creative Commons Attribution (CC BY) license (http://creativecommons.org/licenses/by/4.0/).

Article

Automatic Inundation Mapping Using Sentinel-2 Data Applicable to Both Camargue and Doñana Biosphere Reserves

Georgios A. Kordelas [1], Ioannis Manakos [1,*], Gaëtan Lefebvre [2] and Brigitte Poulin [2]

[1] Information Technologies Institute, Centre for Research and Technology Hellas, Charilaou-Thermi Rd. 6th km, 57001 Thessaloniki, Greece; kordelas@iti.gr
[2] Tour du Valat Research Institute for the conservation of Mediterranean wetlands, Le Sambuc, 13200 Arles, France; lefebvre@tourduvalat.org (G.L.); poulin@tourduvalat.org (B.P.)
* Correspondence: imanakos@iti.gr; Tel.: +30-2311-257760

Received: 19 July 2019; Accepted: 25 September 2019; Published: 27 September 2019

Abstract: Flooding periodicity is crucial for biomass production and ecosystem functions in wetland areas. Local monitoring networks may be enriched by spaceborne derived products with a temporal resolution of a few days. Unsupervised computer vision techniques are preferred, since human interference and the use of training data may be kept to a minimum. Recently, a novel automatic local thresholding unsupervised methodology for separating inundated areas from non-inundated ones led to successful results for the Doñana Biosphere Reserve. This study examines the applicability of this approach to Camarque Biosphere Reserve, and proposes alternatives to the original approach to enhance accuracy and applicability for both Camargue and Doñana wetlands in a scientific quest for methods that may serve accurately biomes at both protected areas. In particular, it examines alternative inputs for automatically estimating thresholds while applying various algorithms for estimating the splitting thresholds. Reference maps for Camargue are provided by local authorities, and generated using Sentinel-2 Band 8A (NIR) and Band 12 (SWIR-2). The alternative approaches examined led to high inundation mapping accuracy. In particular, for the Camargue study area and 39 different dates, the alternative approach with the highest overall Kappa coefficient is 0.84, while, for the Doñana Biosphere Reserve and Doñana marshland (a subset of Doñana Reserve) and 7 different dates, is 0.85 and 0.94, respectively. Moreover, there are alternative approaches with high overall Kappa for all areas, i.e., at 0.79 for Camargue, over 0.91 for Doñana marshland, and over 0.82 for Doñana Reserve. Additionally, this study identifies the alternative approaches that perform better when the study area is extensively covered by temporary flooded and emergent vegetation areas (i.e., Camargue Reserve and Doñana marshland) or when it contains a large percentage of dry areas (i.e., Doñana Reserve). The development of credible automatic thresholding techniques that can be applied to different wetlands could lead to a higher degree of automation for map production, while enhancing service utilization by non-trained personnel.

Keywords: inundation mapping; flood mapping; automatic thresholding; Sentinel-2; wetlands; marshland; Camargue; Doñana

1. Introduction

Wetlands, which constitute unique habitats for many different plant and animal species, are important for their water-related ecosystem services, such as food provision, water filtration, and protection against soil erosion [1]. Additionally, they provide important recreational and leisure activities, such as bird watching, fishing, and hiking [2]. Wetlands are in danger of rapid decline in both quantity and quality due to impacts related to climate change and human pressures [3]. Therefore, monitoring the variability of the surface water extent across time is important for taking actions to increase resilience. Satellite data can provide a cost-effective solution for frequent and accurate monitoring of surface water extent.

Numerous approaches, utilizing optical or radar data for estimating water surface areas, have been proposed in the literature. The advantage of radar-based approaches is that they can operate under nearly all-weather and day-night conditions. However, emergent vegetation [4], waves [5,6], sand [7], and radar shadows produced by terrain features [8] impede the efficient delineation between water and land. On the other hand, the extraction of the water surface from optical imagery is generally more straightforward than radar imagery [9,10], since the rich spectral information of optical data allows for the reliable detection of the water presence by utilizing various indices and bands. The main limitation of using optical data is cloud presence, which prohibits the observation of the earth's surface [10]. Several approaches utilize both optical and radar data to deal with the lack of optical data during extended periods of cloud cover and overcome the limitations of radar data [11,12]. The methodology presented in this work relies on optical data. Thus, literature information focuses on this category of approaches.

Thresholding approaches detect water-covered areas by applying thresholds to one or more spectral bands or indices [13–19]. Commonly used indices include the Normalized Difference Water Index (NDWI) [15,16,19], Modified NDWI [14,18–20], and Automated Water Extraction Index [17–19,21]. Several approaches use information from Shortwave infrared (SWIR) spectral ranges to identify shallow inundated wetland areas, since it is less sensitive to sediment-filled waters and, hence, more efficient for registering the boundaries between water and dry areas in shallow wetlands [13,22–24]. Automatic thresholding approaches can be applied to different areas and are computationally inexpensive, but they may wrongly classify dark objects (i.e., shadows and buildings) as water when their spectral characteristics are similar [25]. Automatic thresholding approaches are distinguished into: (a) global approaches [15,17,19,20,26], which estimate thresholds based on the histogram analysis of the complete image, and (b) local thresholding approaches [23], which estimate local thresholds for image subsets containing high percentages of pixels belonging to the water and non-water classes, and then may take into consideration subsets' thresholds to estimate an overall threshold. Local thresholding approaches overcome the incapability of global approaches to estimate an optimal histogram threshold when the class proportions within the image are imbalanced [27]. Various algorithms (such as Otsu's [28], and Kittler and Illingworth's [29]) have been used for estimating thresholds separating inundated and non-inundated pixels inside an image.

Machine-learning algorithms, including supervised and unsupervised ones, have also been used for detecting water bodies from multispectral imagery. Supervised approaches applied for water detection include random forests [30,31], neural networks [32], support vector machines [33,34], deep neural networks [35,36], and decision trees [13,22,37], while frequently used unsupervised approaches for water mapping include K-means [38] and Iterative Self-Organizing Data Analysis Technique (ISODATA) [39]. Even though machine-learning approaches usually exhibit higher accuracy than thresholding ones [30], they have several limitations. (a) In case reference data is not available, supervised approaches require collection of training samples, which is a time-consuming and tedious task that requires expert knowledge and/or validation in the field [25,40]. (b) Supervised methods may meet problems when mapping water bodies over large scale areas [41]. (c) Unsupervised methods need expert knowledge to select the initial class and iteration parameters [42], and may need post processing of the results to combine adjacent regions into larger regions corresponding to water bodies [30].

Most of the previously mentioned approaches aim to detect water in open-water bodies such as rivers, lakes, reservoirs, and watersheds [17,19,20,30,33,37,43], while a part of them focuses on wetland areas [13,22–24].

This study focuses on wetland areas and examines alternative approaches of the original automated local thresholding approach presented in Reference [23] with the objective to suggest alternative approaches, which may produce credible results for both Camargue and Doñana wetlands, i.e., be identified as possibly applicable to further wetland areas and biomes. Each examined alternative approach relies on a specific band or band combination, acknowledged as effective by the underlying physics, and a specific approach for estimating splitting thresholds. The different Sentinel-2 (S2) based inputs examined for estimating thresholds include: (i) Band 11 (SWIR-1), (ii) product (result of the multiplication) of Band 12 (SWIR-2) and Band 8A (NIR), and (iii) product of SWIR-1 and NIR. The different methods for estimating splitting thresholds include: (i) minimum entropy thresholding, and (ii) Otsu's algorithm. The results of the alternative approaches are compared against reference maps, provided for Doñana and Camargue by local research institutes, based on locally developed water detection models [22,24].

2. Materials and Methods

2.1. Study Areas

2.1.1. Camargue

The Camargue (Figure 1) is a polderized delta created by the Rhône River that covers an area of 145,300 ha rarely exceeding 5 m in altitude. It comprises a high diversity of wetland types according to a water-salinity gradient such as lagoons, salt meadows, dense halophilous scrubs and steppes, brackish/freshwater marshes with tall emergent or permanent aquatic vegetation, and temporary pools. The climate is Mediterranean, being characterized by mild and wet winters and hot and dry summers. With mean annual precipitations of 600 mm, mainly concentrated from autumn to early spring, and a mean evapotranspiration of 1400 mm, the Camargue is characterized by high water deficits, especially in the summer [44,45]. As a result, 730 millions of cubic meters of water are pumped from the Rhône on average each year to compensate for river embankment, avoid soil salinization, and enhance primary production. This water, primarily pumped from March to September, is distributed through a complex network of channels for irrigating crops and pastures, as well as for flooding marshes, which support nature conservation, wildfowl hunting, and reed harvest. Accordingly, some Camargue wetlands are flooded year round or during most of the year as a result of artificial irrigation (e.g., hunting reed marsh). Other wetlands become naturally dry during the summer season (e.g., harvested reed marsh, bulrush marsh) and, lastly, a few are flooded only during a period of high rainfalls (e.g., halophilous scrubs, temporary pools) [46]. Within the context of climate change and increasing human pressures, the development of a remotely-sensed tool to monitor water seasonal variation in these wetlands is essential [47].

2.1.2. Doñana

The Doñana wetlands (Figure 2), covering 108,429 ha, lie within and around the delta of the Guadalquivir River in Southwest Spain and contain two main habitat types: seasonal marshes and adjacent eolian sands holding aquifer-fed dune ponds. These are surrounded by scrublands, pine forests, and cultivated areas. The marshland area is comprised of seasonal open marsh with emergent plants, temporary pools with annual plant species, and scattered halophilous scrubs [48]. The Doñana climate is Mediterranean sub-humid with hot and dry summers and mild and wet winters. The annual precipitation, with an average of 550 mm, occurs mainly between October and April and is almost absent between May and September. The highest monthly rainfall usually occurs in November and maximum water levels are reached during February. Marshes dry up slowly in late spring and most of their surface gets completely dry by the end of July [49]. Marshland's depth, turbidity, and vegetation cover varies depending on the amount and seasonal pattern of precipitation [24]. Marshes are breeding ground and host many species of migratory birds during the winter [49]. In parallel, rice-paddies, aquaculture ponds, and salt pans constitute an extensive system of artificial wetlands in Doñana that also provide a habitat for water birds and other species, especially when natural wetlands dry up seasonally [50].

Remote Sens. **2019**, *11*, 2251

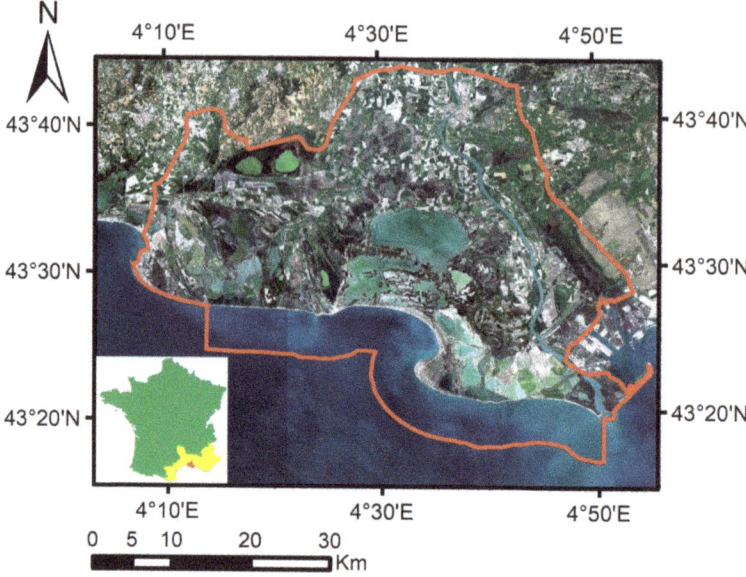

Figure 1. Map of the Camargue Biosphere Reserve located in south France, partially falling within the Occitanie and Provence-Alpes-Côte d'Azur administrative regions, with underlying S2 RGB (red-green-blue) image on 18 April 2018. Red line: boundary of the Biosphere Reserve.

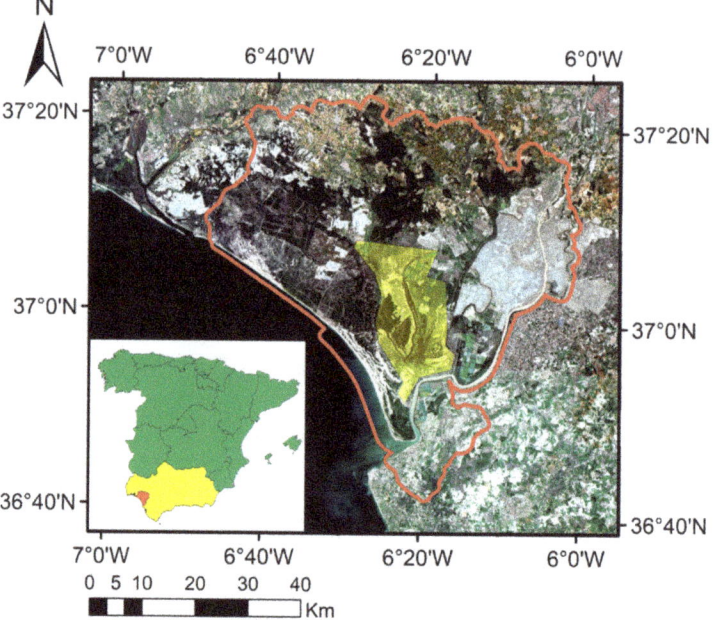

Figure 2. Map of the Doñana Biosphere Reserve located in Southwest Spain, falling within the Andalusia administrative region, with underlying S2 RGB image on 21 February 2018. Red line: boundary of the Biosphere Reserve. Yellow shaded area: Doñana marshland wetland area.

2.2. Dataset

2.2.1. Camargue

Thirty-nine cloud-free and atmospherically corrected S2 Level-2A (L2A) products of the Camargue from 12 June, 2017 to 19 June, 2018 were downloaded from the Copernicus European Space Agency (ESA) hub (see Table 1). The shapefile containing the boundaries of Camargue comprises two different tiles (namely 31TEJ, 31TFJ). The tile products are mosaicked and clipped to the extent of the shapefile.

Table 1. Dates of cloud free S2 acquisitions over Camargue.

Cycle	Sept.	Oct.	Nov.	Dec.	Jan.	Feb.	Mar.	Apr.	May	June	July	Aug.
2016–2017										12,19	4,7 12,14 17	3,13 18,21
2017–2018	5,7 20,27	5,7 10,12 27,30	14,16 19,21	6,9 16,24	23	4,22 27	14	18,20	20,25	19		

The reference maps for Camargue were obtained by dichotomous partitioning of reflectance values encoded as 1 for water presence and 0 for water absence based on ground-truth (n = 1229) and optical-space derived (n = 2603) reference points covering the whole Biosphere Reserve area and all the main habitat types [22]. Ground-truth data refer to water level measures in different wetland types, focusing on those with a dense vegetation cover. For optical-space data, a water formula developed with SPOT-5 (Satellite Pour l'Observation de la Terre - 5) [13] was used to generate a water mask for dates when Landsat-8 scenes were also available. Fifty random points from each of the main land cover types in the Camargue (n = 17) were selected on the SPOT-5 water maps and transferred to Landsat-8 scenes with similar dates. These points were then used as training data along with ground-truth measures for creating a Landsat-8 water formula. The same procedure was repeated to develop a water formula with S2 from Landsat-8. Data mining was performed with the Rpart package in the R software using eight bands of S2, as well as various spectral indices found in the literature for explaining the variables (for more details, see References [13,22]). A random selection of 30% of all points was excluded from the sample and used for validation. The best model selected used the near (NIR) and short-wave (SWIR-2) infrared wavelengths. NIR was useful for discriminating areas that are completely dry, while SWIR-2 was efficient for water detection. This model provided an overall accuracy of 94% for predicting water presence/absence with Kappa coefficients of 0.82 on both the training and validation samples [22]. Wetlands characterized by a dense cover of vegetation were correctly classified at 89%.

2.2.2. Doñana

Seven cloud-free and atmospherically corrected S2 L2A products of the Doñana from 1 June, 2017 to 17 April, 2018 were downloaded from the ESA hub, so that they timely overlap with Landsat cloud-free parallel data (see Table 2). The shapefile containing the boundaries of Doñana Biosphere Reserve comprises three different tiles (namely 29SQA, 29SQB, and 29SPB). The tile products were mosaicked and clipped to the extent of the shapefile.

Table 2. Dates of cloud free S2 acquisitions over Doñana, coinciding with Landsat ones, with the exception of one date. Landsat acquisition was one day earlier than the S2 acquisition on 21 February, 2018.

Date	01/06/2017	11/07/2017	20/08/2017	08/11/2017	27/01/2018	21/02/2018	17/04/2018

Inundation maps of the Doñana area with 30-m pixel resolution, which are generated from Landsat satellite data and provided by the Doñana Biological Station (EBD), are used as ground truth data.

These reference maps were obtained by dichotomous partitioning of reflectance values on Landsat Thematic Mapper (TM) and Enhanced TM (ETM) band 5 (SWIR) based on 6005 ground-truth field sampling points, mostly collected within the marshland area of Doñana [24]. This model provided an overall accuracy of 93% for predicting water presence/absence with a Kappa coefficient of 0.65.

2.3. Methodology

The work presented in Reference [23] introduces an unsupervised approach, detecting automatically thresholds on the SWIR-1 band and on a Modified-Normalized Difference Vegetation Index (MNDVI) (estimated as the normalized difference of Band 7 and Band 5 of S2), to estimate open-water and water-vegetation subclasses. This approach demonstrated high classification accuracy for Doñana, with an overall Kappa coefficient reaching 0.94 and 0.88 for the marshland and the complete area (i.e., the Biosphere Reserve), respectively. This paper examines alternatives of the original thresholding approach driven by the findings for Camargue. Instead of the SWIR-1 band, other algebraic band combinations are also examined. Additionally, both the minimum cross entropy thresholding algorithm (MCET) [51] and Otsu's algorithm [28] are used for estimating thresholds partitioning inundated and non-inundated pixels inside an image subset based on their class distribution. The methodology's steps are presented in Figure 3.

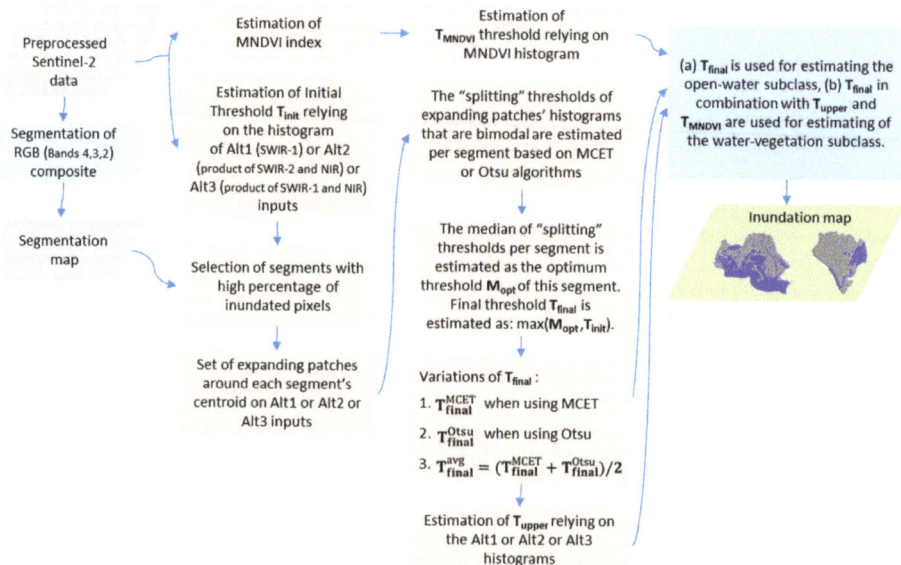

Figure 3. Schematic flow diagram of the automatic thresholding methodology.

2.3.1. Segmentation of the Satellite Image

As the first step, the satellite image is segmented into non-overlapping regions by utilizing the mean-shift segmentation algorithm [52]. The resulting segmentation map is utilized for selecting segments with a high percentage of inundated pixels.

The input to the segmentation algorithm is a false color image composed of Band 2 (BLUE), Band 3 (GREEN), and Band 4 (RED), which have been normalized to the range [0,255], by relying on minimum and maximum values representing the intensity percentile range from 1% to 99% per band. These bands are selected due to their 10-m resolution, which allows for a more accurate segmentation of the satellite image, compared to selection of other bands with a lower resolution [23].

2.3.2. Mapping of the Open-Water Subclass

Three input data alternatives are examined for estimating an initial threshold T_{init} separating inundated from non-inundated pixels: (i) SWIR-1 band, which is denoted as Alt1, (ii) product (per pixel multiplication) of SWIR-2 and NIR, which is denoted as Alt2, and (iii) product of SWIR-1 band and NIR, which is denoted as Alt3. The first alternative Alt1 is the one proposed in the original approach [23]. The second alternative Alt2 is inspired by the approach suggested in Reference [22] for Camargue, which applies strict thresholds to SWIR-2 and NIR (see paragraph 2.2.1), and the third alternative Alt3 is a variant of the Atl2 where SWIR-1 is used instead of SWIR-2. Each of Alt1, Alt2, and Alt3 is normalized in the range of [0,255], by relying on minimum and maximum values representing the intensity percentile range from 1% to 99%. The term **"Inp"**, used in the following, is a generic term for the input data, corresponding to either Alt1, Alt2, or Alt3.

The histogram of **Inp** is used to estimate an initial threshold T_{init} separating inundated from non-inundated pixels. T_{init} is the **Inp** value for which the first deep valley of this histogram is detected (Figure 4 shows histograms using Alt2 as input). If a pixel **p** has $Inp(p) < T_{init}$, it is denoted as inundated. Otherwise, it is denoted as non-inundated. Therefore, an initial inundation map is generated. Based on this inundation map, segments G_m, where $m = 1, 2, \ldots, M$, having a large percentage of inundated pixels (i.e., over 70% of a segment's pixels are inundated) are selected and their corresponding centroids C_m are estimated.

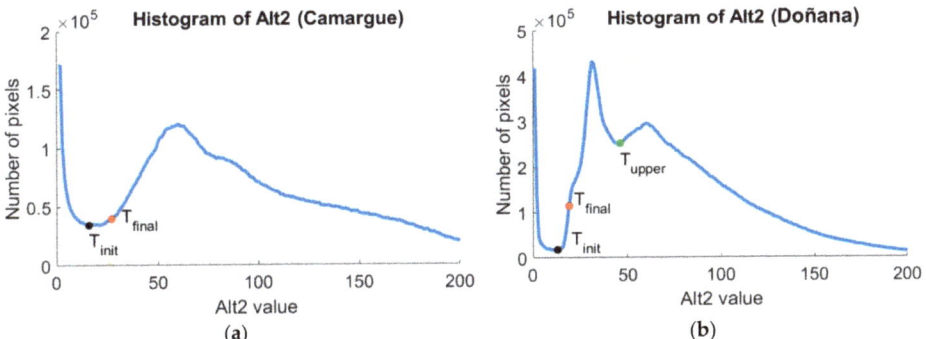

Figure 4. Histogram of the Alt2 map for (**a**) Camargue on 18 August 2017, and (**b**) Doñana on 20 August 2017. T_{init} is the value for which the first deep valley in the Alt2 histogram is detected (black point). The red point on the Alt2 histogram corresponds to the final Alt2 threshold T_{final} estimated by relying on MCET. The deep valley right after the first valley in the Alt2 histogram, detected only on the right histogram, corresponds to threshold T_{upper} (green point). Histogram values reach up to 255 normalized intensity value. However, since there are a few pixels with an Alt2 value over 200 and they do not cause fluctuations of the Alt2 histogram curve, the upper limit of the x-axis is set to 200 for visualization reasons.

Square patches P_m^k of expanding window size are centered around each C_m. The window size in pixels is given by $[20 \cdot k \times 20 \cdot k]$, where $k = 1, 2, \ldots, 20$. The splitting thresholds f_m^k of patches with bimodal histogram (i.e., two distinctive classes of intensity values corresponding to inundated and non-inundated classes appear in the histogram) are estimated based on the histogram thresholding algorithms (MCET or Otsu's algorithm). Their median is assumed to be the optimal threshold f_{opt}^m corresponding to the segment G_m. The usefulness of using expanding patches is to calculate the optimal threshold per segment more robustly, since its estimation is based on multiple splitting thresholds. Then, the median of selected segments' optimal thresholds is estimated as: $M_{opt} = \underset{m=1,2,\ldots,M}{\text{median}} (f_{opt}^m)$.

Final threshold T_{final} (see Figure 4), discriminating the open-water subclass, is estimated as $\max(M_{opt}, T_{init})$. Pixels **p** with $Inp(p) < T_{final}$ are assumed to belong to the open-water subclass, comprising of water or water with sparse vegetation pixels.

When MCET or Otsu is used for estimating the splitting thresholds, T_{final} is denoted as T_{final}^{MCET} or T_{final}^{Otsu}, respectively. Average (abbreviated as "avg"): $T_{final}^{avg} = \left(T_{final}^{MCET} + T_{final}^{Otsu}\right)/2$ is also examined as the final threshold.

2.3.3. Mapping of the Water-Vegetation Subclass

Inundated areas may be covered by emergent vegetation. The **Inp** value of the pixels in these areas is higher compared to the **Inp** values of the pixels with open water or water with sparse vegetation belonging to the open-water class. Therefore, a threshold T_{upper}, which is the **Inp** value for which the deep valley right after the first deep valley is detected in the **Inp** histogram, is assumed to represent the upper threshold for pixels comprising areas of water covered by dense vegetation (see Figure 4b). At the same time, areas with dense emergent vegetation are expected to have high MNDVI value. Thus, a threshold T_{MNDVI}, which should be more than 0.4, is selected. In particular, T_{MNDVI} is set equal to the MNDVI value, for which the first valley in the part of the histogram with MNDVI values over 0.4 is detected (see Figure 5b) following the findings in Reference [23].

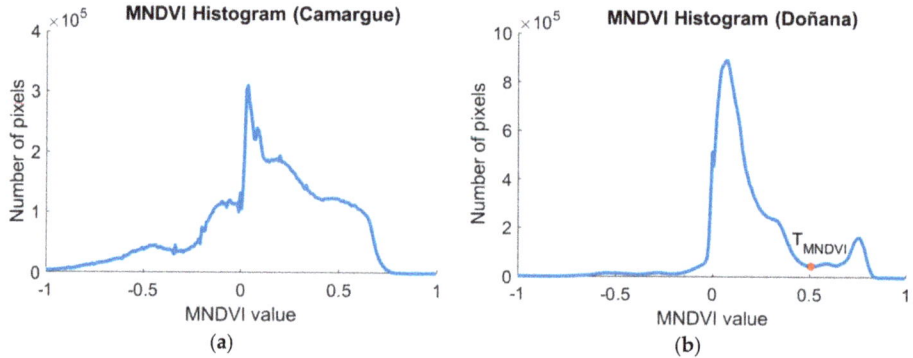

Figure 5. Histogram of the MNDVI map for (**a**) Camargue on 18 August, 2017, and (**b**) Doñana on 20 August, 2017. T_{MNDVI} is the MNDVI value for which the first deep valley after 0.4 is detected in the MNDVI histogram (red point). T_{MNDVI} is detected only on the right histogram.

A pixel **p** is assumed to belong to the water-vegetation subclass if the following two conditions are met: $T_{final} < Inp(p) < T_{upper}$ & $MNDVI(p) > T_{MNDVI}$, presuming that T_{upper} and T_{MNDVI} can be detected in the histograms of **Inp** and MNDVI, respectively. These latter conditions are valid for Doñana on 20 August, 2017 (see Figures 4b and 5b), but not for Camargue on 18 August, 2017 (see Figures 4a and 5a).

3. Results

3.1. Comparison of Automatic Thresholding Results Against Reference Map of Camargue

The three input alternatives (Alt1 or Alt2, or Alt3) and the three different ways of estimating the final threshold (T_{final}^{MCET} or T_{final}^{Otsu} or T_{final}^{avg}) form nine different alternatives, with each one combining an input alternative and a way of estimating the final threshold. The alternative ways of estimating final threshold are referred to as "MCET" (standing for T_{final}^{MCET}), "OTSU" (standing for T_{final}^{Otsu}), and "avg" (standing for T_{final}^{avg}). For example, if "Alt2" is the input and "avg" is the way of estimating the final threshold, then the alternative will be named as "Alt2 (avg)".

Regarding Camargue, each of the alternatives is applied to the 39 S2 products (see Table 1), and the results are compared to the reference maps. Pixels corresponding to the coastal waters are not taken into consideration for the comparison to obtain unbiased classification results. The results are presented using Kappa coefficient, which is estimated in terms of the observed agreement (p_o) and the agreement expected by chance (p_e) as: Kappa= (p_o − p_e) / (1-p_e) [53].

The curves presented in Figure 6 show how the Kappa coefficient varies for the time period between 12 June, 2017 and 19 June, 2018 for five out of nine alternatives with the best agreement with the reference map. The best agreement is given by "Alt3 (OTSU)". "Alt3 (avg)" and "Alt2 (avg)" rank second and third, respectively, and have similar results. Lastly, "Alt3 (MCET)" and "Alt2 (MCET)" rank fourth and fifth, respectively, and give very close results. The previously mentioned ranking is based on the overall Kappa accuracy provided in Section 3.3. For all curves, it is seen that agreement decreases in the winter months from XII to II. For this period, Kappa is below 0.7 for most of the dates, while, for the rest of the year, the Kappa value is generally more than 0.7.

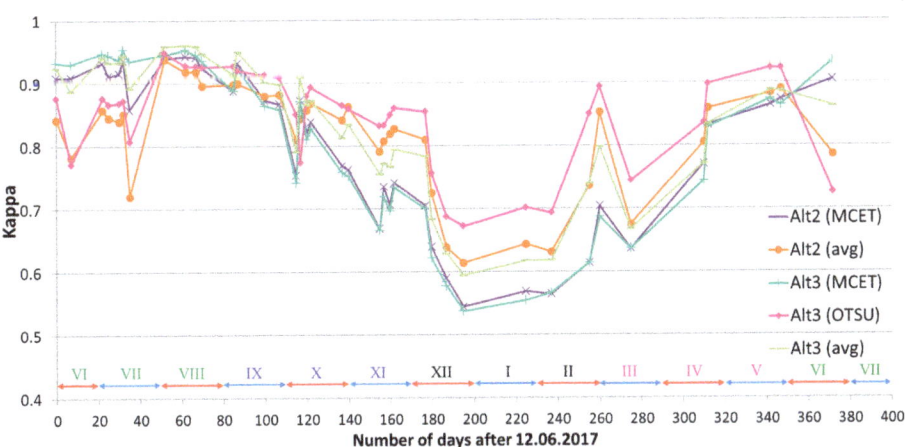

Figure 6. Kappa coefficient variation of five alternatives in the time period between 12 June, 2017 and 19 June, 2018. The months' numbers are given as Latin numerals and are colored according to the season they belong, i.e., green for summer, blue for autumn, black for winter, and pink for spring.

Figure 7 shows an example of inundation maps obtained from an S2 scene on 27 February, 2018 when using the alternative approaches of Alt3 input as compared to the reference map. Green and pink colors indicate classification differences between the reference map and the Alt3 maps. The map of "Alt3 (OTSU)" (Figure 7b) agrees better with the reference map (Figure 7a), compared to "Alt3 (MCET)" (Figure 7d), which largely underestimates water presence, and "Alt3 (avg)" (Figure 7c), which shows intermediate results. However, while "Atl3 (OTSU)" is able to detect water in vegetated areas more efficiently, such as the ones enclosed in the yellow squares, it tends to overestimate water presence in agricultural areas.

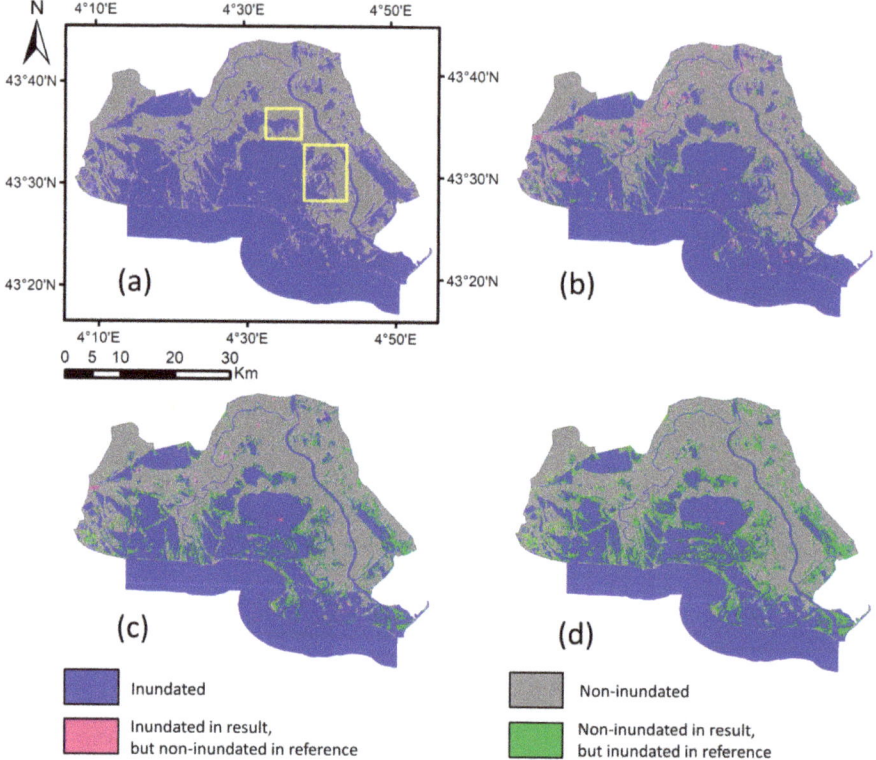

Figure 7. Inundation map on 27 February, 2018 using (**a**) an S2 derived inundation reference map, (**b**) "Alt3 (OTSU)", (**c**) "Alt3 (avg)", and (**d**) "Atl3 (MCET)". Blue and gray colors correspond to inundated and non-inundated areas, respectively. The green color corresponds to areas classified as inundated in (**a**) but non-inundated in (**b**–**d**), while pink color corresponds to areas classified as non-inundated in (**a**) but inundated in (**b**–**d**). The yellow colored squares in (**a**) enclose emergent vegetated areas. The upper yellow square encloses reed beds and the lower one encloses short helophytes, submerged macrophytes, halophilous scrubs, and annual and perennial herbs.

3.2. Comparison of Automatic and Thresholding Results Against Landsat Reference Maps of Doñana

In order to account for the uncertainty caused by the lower spatial resolution of the Landsat-derived reference maps, pixels in the transition zones between inundated and non-inundated areas (i.e., pixels including in their eight closest-neighbor pixels at least one pixel of a different class) in the S2 maps were excluded from the accuracy estimation (see Reference [23] for more information). Pixels in the area corresponding to sea coastal waters were also excluded. The curves, presented in Figure 8, show how Kappa coefficient varies for seven different dates regarding the five top ranking alternatives at the Doñana marshland (Figure 8a) and Doñana complete area (Figure 8b).

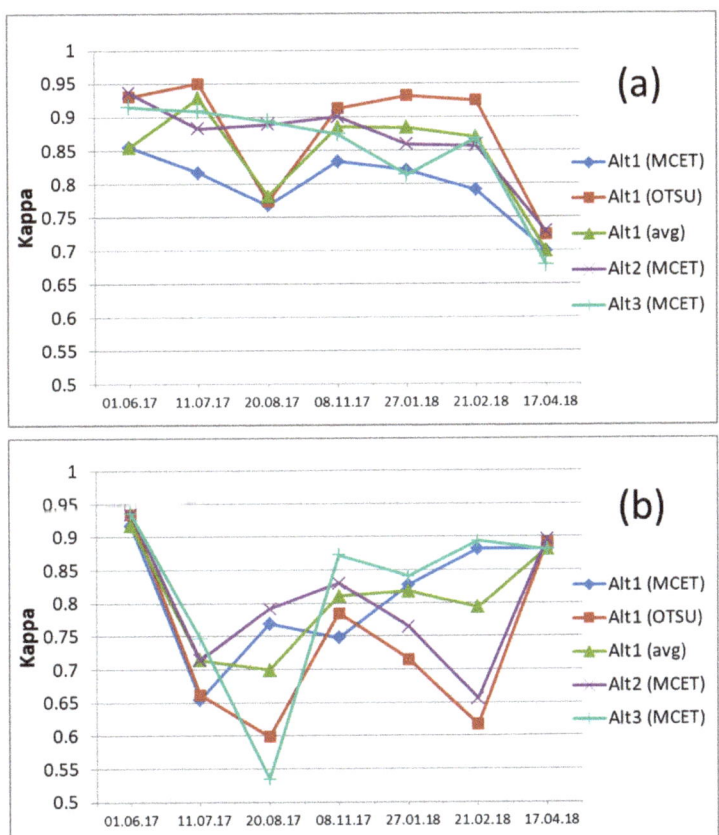

Figure 8. Kappa coefficient variation of five alternatives for seven different dates with respect to (**a**) Doñana marshland and (**b**) Doñana complete area.

In Doñana marshland, "Alt1 (OTSU)" provides the highest accuracy, followed by "Alt2 (MCET)" and "Alt1 (Avg)" (see Section 3.3 for more details on the ranking that is based on overall Kappa).

In the Doñana complete area, "Alt3 (MCET)" provides the highest accuracy, followed by "Alt1 (MCET)" and "Alt2 (MCET)" (see Section 3.3 for more details on the ranking that is based on overall Kappa). "Alt1 (avg)" provides the most consistent Kappa values across the study period.

Figure 9 shows an example of the Landsat derived inundation reference map on 27 January, 2018 and the inundation maps estimated based on the alternative approaches using Alt3 input. Green and pink colors indicate the classification differences between the reference map and the alternative maps. "Alt3 (MCET)" (Figure 9d) agrees well with the reference map (Figure 9a), while "Alt3 (OTSU)" (Figure 9b) largely overestimates the water presence and "Alt3 (avg)" (Figure 9c) shows intermediate results.

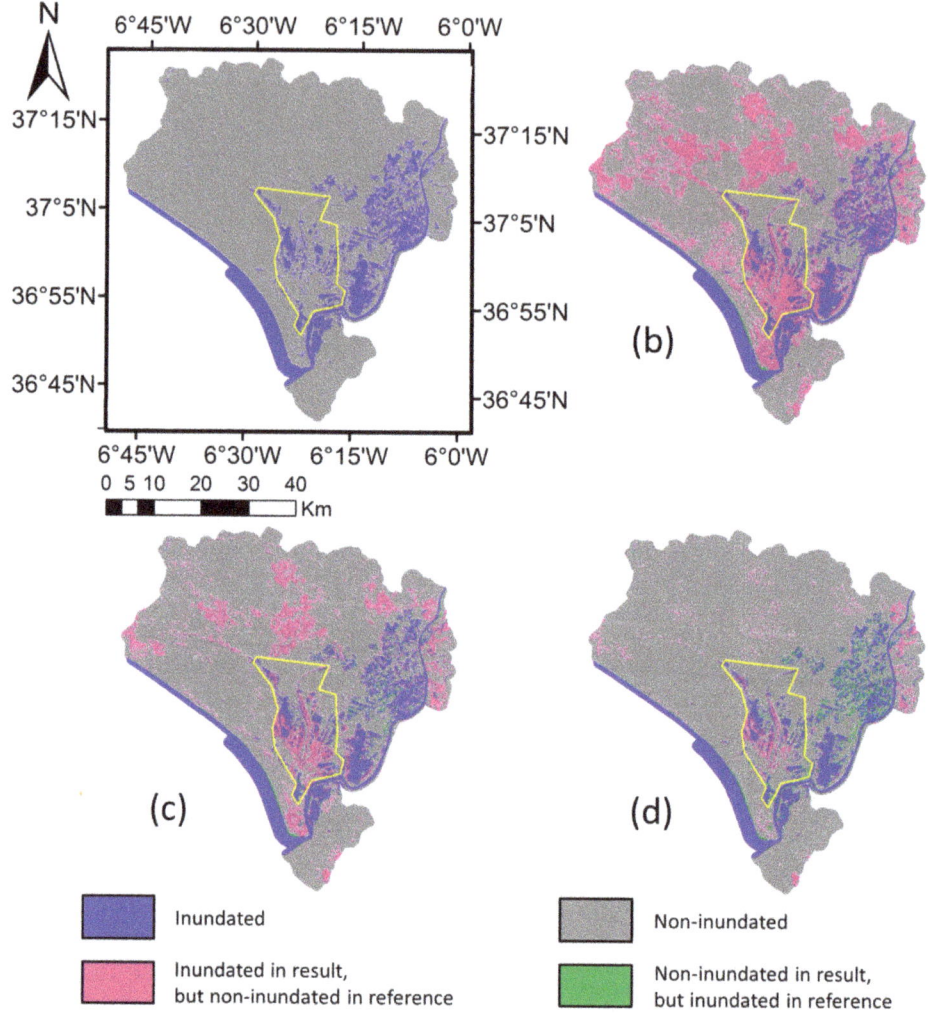

Figure 9. (**a**) Landsat derived inundation map on 27 January, 2018. Inundation map on 27 January, 2018 using (**b**) "Alt3 (OTSU)", (**c**) "Alt3 (avg)", and (**d**) "Atl3 (MCET)". Blue and gray colors correspond to inundated and non-inundated areas, respectively. Green color corresponds to areas classified as inundated in (**a**) but non-inundated in (**b–d**), while the pink color corresponds to areas classified as non-inundated in (**a**) but inundated in (**b–d**). The yellow polygon encloses marshland area.

3.3. Overall Kappa Per Approach and Examined Areas

Figure 10 provides overall Kappa coefficients, which are estimated when the number of true positive pixels of the inundated class, false positive pixels of the inundated class, true positive pixels of the non-inundated class and false positive pixels of the non-inundated class are added for dates per approach and examined study. The combined Overall Accuracy of each approach per study case is also given as a number close the point corresponding to its overall Kappa. Kappa values can be classified according to Reference [53] as follows: "Moderate" when $0.40 < \text{Kappa} \leq 0.60$ (shortened as "Mod"), "substantial" when $0.60 < \text{Kappa} \leq 0.80$ (shortened as "Sub") or "almost perfect" when $0.80 < \text{Kappa} \leq 1$ shortened as ("Alm Perf").

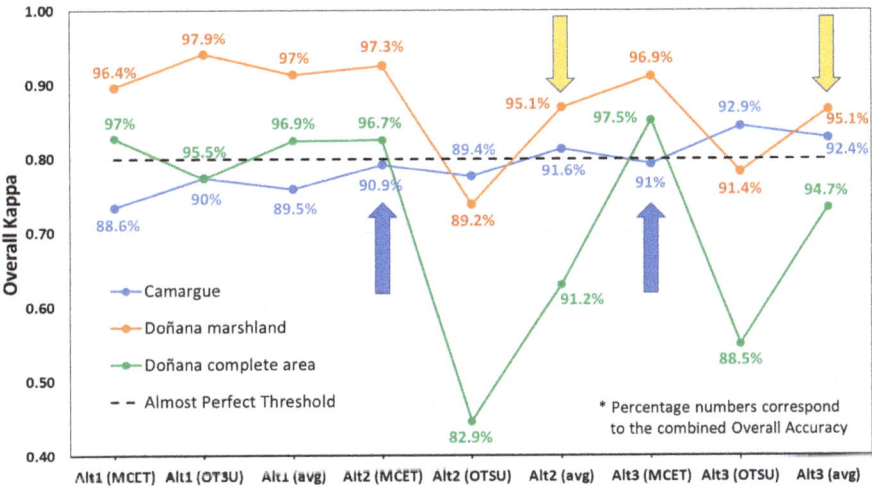

Figure 10. Overall Kappa per approach and examined study area. The dotted line delineates the threshold of the "Alm Perf" case. Yellow arrows indicate alternative approaches succeeding "Alm Perf" results in Camargue and Doñana marshland at the same time, while blue arrows showcase the most successful overall alternative ones. The combined overall accuracy of each approach per study area is given as a number close the point corresponding to its overall Kappa.

"Alt2 (avg)" and "Alt3 (avg)" (see yellow arrows in Figure 10) have both "Alm Perf" overall Kappa when considering Camargue and Doñana marshland study areas. While, for the Doñana complete area, their Kappa is "Sub", since overall Kappa for "Alt2 (avg)" and "Alt3 (avg)" is 0.63 and 0.73, respectively. On the other hand, "Alt2 (MCET)" and "Alt3 (MCET)" (see blue arrows in Figure 10) exhibit more consistent performance across sites, since their Kappa is "Alm Perf" for both Doñana marshland and Doñana complete area, while, for Camargue, their overall Kappa is very close to "Alm Perf."

4. Discussion

The aim of this paper is to examine the performance of various alternative automatic thresholding approaches for inundation mapping in two different wetland areas, which include Camargue and Doñana. These are considered as testbeds to evaluate the transferability of the different approaches and select the ones that favor the acquisition of consistently credible results for both wetlands. In this study, the local thresholding alternative approaches utilize multispectral satellite data, while the vast majority of local thresholding approaches have been designed for utilizing radar data (e.g., [27,54–58]). Moreover, the estimation of local thresholds relies on patches of expanding size, contrary to most of the approaches using image subsets of fixed size [55–58]. This helps to increase robustness, when input data have different spatial resolution such as when inputs from different satellites are used.

The overall Kappa of the alternative approaches varies from 0.73 to 0.84 for Camargue, from 0.74 to 0.94 for Doñana marshland, and from 0.45 to 0.85 for the complete area of Doñana. The experimental results demonstrate that there is not a common approach across all study cases achieving absolute top accuracy. "Alt3 (OTSU)" is best for Camargue, "Alt1 (OTSU)" is best for Doñana marshland, and "Alt3 (MCET)" is best for Doñana as a whole. The use of "Alt2" and "Alt3" input alternatives leads to the estimation of additional inundated areas compared to "Alt1", but sometimes leads to overestimating the water presence in dry areas. "OTSU" algorithm identifies the final threshold at a relatively high value leading to a possible overestimation of the inundated areas, while the "MCET" algorithm identifies a final threshold at a lower value, which leads to a possible underestimation of the inundated areas, especially in emergent vegetation areas. As a sequence, the use of "avg" helps to balance between the overestimation and underestimation of inundated areas. "Alt2 (avg)" and "Alt3 (avg)" are the most accurate approaches when considering, at the same time, Camargue and Doñana marshland, with both approaches exhibiting overall Kappa over 0.81 and 0.86 for the two study areas, respectively. These areas are dominated by emergent vegetation and temporary flooded areas, and, thus, "Alt2 (avg)" and "Alt3 (avg)" could be utilized for other wetlands with similar characteristics. However, the overall Kappa is below 0.74 for both approaches when considering the Doñana complete area. The complete Doñana area is comprised of larger parts that are permanently dry. In this case, the utilization of MCET achieves a higher accuracy. Overall, "Alt2 (MCET)" and "Alt3 (MCET)" showcase a consistent overall Kappa exceeding 0.79 for all three study areas, since they avert the erroneous detection of water in dry areas. Thus, these two alternatives should be preferred for wetlands comprising large parts of dry areas.

The fact that different approaches seem to operate best in one or the other study area may be related to the models used for generating the reference maps, as well as the original field data used to build the local water presence models. In Doñana as in Camargue, the reference maps are from formulas developed to optimize water detection in the wetland types that represent very effective local conditions. Doñana marshland mostly includes seasonal marshes with relative sparse and low emergent vegetation due to the short flooding period. Tall emergent plants are rare and they are generally grazed by cattle when present [48]. Hence, the formula originally developed in Doñana does not have to perform well under dense vegetation cover. On the other hand, tall emergent plants cover large areas of semi-permanent marshes in Camargue, and the formula originally developed was specifically meant to detect water under dense and tall vegetation cover [22]. The sensors and spectral bands used to develop the original formula at each site could also influence the performance of each alternative approach tested. In Doñana, the estimation of the reference maps relied on band 5 of Landsat TM and ETM, which is similar to the SWIR-1 band of S2. The best performing alternative at this site is associated with the SWIR-1 band (e.g., Alt1 data input). In Camargue, the original formula used a combination of SWIR-2 and NIR S2 bands [22], and the three best performing alternatives use NIR in combination with SWIR-1 (e.g., Alt2 data input) or SWIR-2 (e.g., Alt3 data input) bands.

In Camargue, the accuracy of all alternative approaches decreases during winter (from December to February). This systematic error is largely due to the misclassification of the water presence under dense cover of scrubby vegetation (Salicornia marshes). This habitat, which is flooded by rainfall during this specific time of year [22], contributes to 25% of water pixels that were misclassified as dry in Figure 7. Another example is given in Figure 11, where the comparison between (a) and (b) shows that "Alt3 (OTSU)" mainly detects open water areas, and neglects the water presence under Salicornia salt marshes classified as flooded in the reference map.

During the study period, S2 satellites were passing over Camargue during late morning (between 10.10 and 10.40 CET). As a consequence, shadows, which are mainly observed in the town of Arles in the north of Camargue area, appear in S2 data. Longer shadows, which are observed from the end of autumn to early spring and mainly during winter, are misclassified as inundated areas to a larger degree with the reference map compared to the automatic thresholding alternatives. Therefore, shadow misclassification related to time of the satellite passage factor can also contribute to the reduction of agreement between the reference map and automatic thresholding approaches in the winter. Another example of disagreement is related to deep waters in a large lagoon that were misclassified as non-inundated areas by the reference map in November (14 November 2017) but not by "Alt2 (avg)" (Figure 12). This misclassification (see within the red square of Figure 12b) was attributed to the presence of waves caused by strong winds during the S2 image acquisition. Presumably, other sources of "false misclassifications" could arise from the reference maps of Doñana and Camargue because they were built from models that were not 100% accurate.

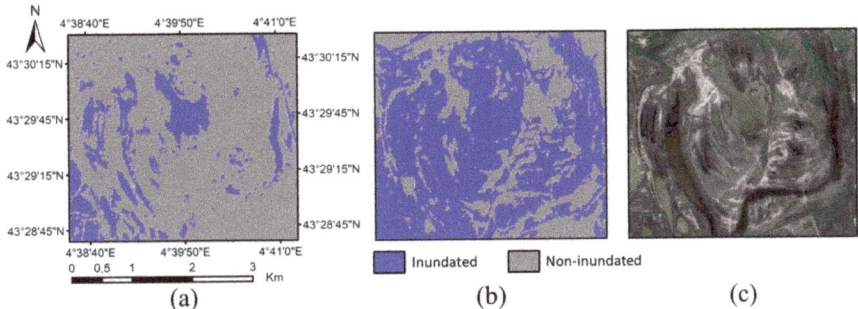

Figure 11. Zoomed part of the inundation map generated on 23 January, 2018 using: (**a**) "Alt3 (OTSU)", and (**b**) reference map. In this region of the Camargue, the main habitat is halophilous scrubs (Salicornia marshes). Areas of open water and bare ground are clearly visible on the Google Earth image of the same area acquired on 22 April, 2018 shown in (**c**).

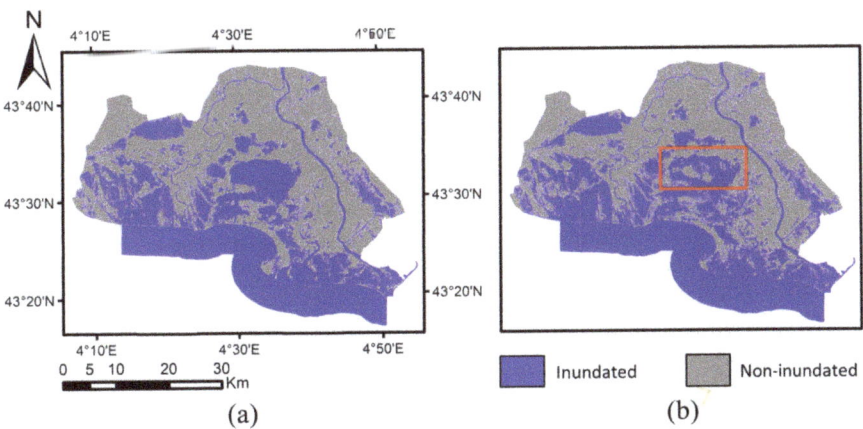

Figure 12. Inundation map generated on 14 November, 2017 using: (**a**) "Alt2 (agv)", and (**b**) reference map. Blue and gray colors correspond to inundated and non-inundated areas, respectively.

Another finding is related with the rice fields within the study areas. The criteria for mapping the water-vegetation subclass (see paragraph 2.4.3) as defined for Doñana are not satisfied for Camargue for any of the dates or any of the alternative approaches. On the contrary, for Doñana, the criteria are satisfied for the dates falling within summer (e.g., 11 July, 2017 and 20 August, 2017) and the water-vegetation subclass is mainly detected in the upper east area where rice fields are located. This result complies with the results in Reference [23] and is in agreement with the growing cycle of rice, since the rice grows during the summer period in Doñana and, at the same time, the paddies are flooded [59]. The difference between Camargue and Doñana is that the thresholds examined in the criteria for detecting water-vegetation subclass can be detected in the **Inp** for Alt1 and Alt2 and MNDVI histograms of Doñana, but not of Camargue (see Figures 4 and 5). This likely relates to the land cover synthesis of each area. In particular, for the rice paddies of Camargue, it is found that there are other land cover types (e.g., reed marsh) with similar spectral behavior (i.e., similar **Inp** and MNDVI values) to the rice paddies. This is evident in a much smaller degree for Doñana. Moreover, the average MNDVI value of the Camargue rice paddies is lower than the average MNDVI value of the Doñana rice paddies and, thus, the discrimination of Camargue rice paddies from other vegetated areas in the MNDVI histogram is impeded. Due to the previously mentioned reasons, emergent rice cannot be detected in Camargue from the middle of July to early September.

Furthermore, by comparing Alt1, Alt2, and Alt3, it is evident that Alt2 and Alt1 histograms allow the detection of T_{upper} (see Figures 4b and 13a), while, in the histogram of Alt3, T_{upper} cannot be detected (see Figure 13b). As a consequence, when comparing the inundation maps presented in Figure 14, it is evident that the water-vegetation subclass, corresponding mainly to rice fields located in the upper right part of the inundation map, cannot be detected when using the "Alt3 (MCET)" approach.

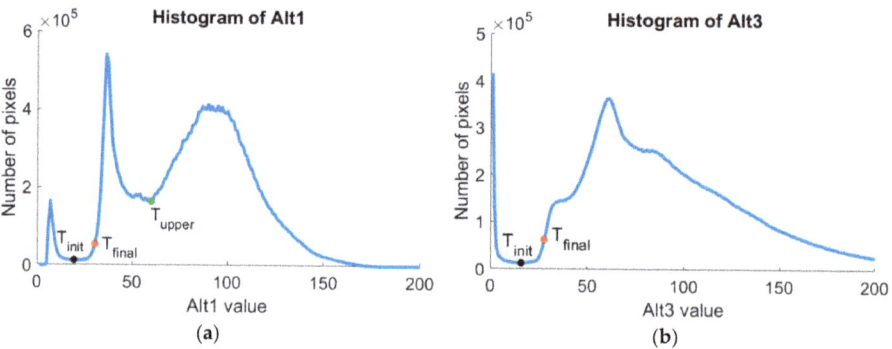

Figure 13. Histogram of the (**a**) Alt1 and (**b**) Alt3 maps for Doñana on 20 August, 2017. T_{init} is the value for which the first deep valley in the Alt1 and Alt3 histograms is detected (black point). The red point on Alt1 and Alt3 histograms corresponds to T_{final} estimated relying on MCET. The deep valley right after the first valley in the Alt1 histogram corresponds to threshold T_{upper}. T_{upper} cannot be detected on the Alt3 histogram. (Histograms' values reach up to 255. However, since there are a few pixels with value over 200 that do not cause fluctuations of the histograms' curves, the upper limit of the x-axis is set to 200 for visualization reasons.).

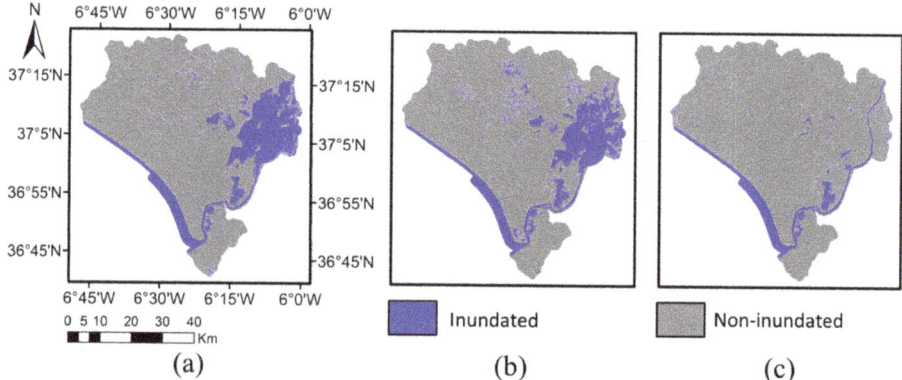

Figure 14. Inundation map generated on 20 August, 2017 using: (**a**) "Alt1 (MCET)", (**b**) "Alt2 (MCET)", and (**c**) "Alt3 (MCET)" alternatives. Blue and gray colors correspond to inundated and non-inundated areas, respectively.

This study proves that automatic thresholding can be applied to more than one study areas and achieve high inundation mapping accuracy, without the need for simultaneous ground truth data or user's intervention. Machine learning approaches that are developed based on ground truth data derived from a specific site [13,30,33,35,37] may perform very accurately as well. However, a credible performance for other sites cannot be safeguarded. Several studies have identified that surface reflectance accuracy of S2 L2A products may vary for different dates [60,61]. Hence, the performance of machine learning approaches could be negatively affected for dates that there is a notable variance in the surface reflectance accuracy compared to the dates, for which the training samples were derived in order to train the classification models.

5. Conclusions

This study examines automatic thresholding alternative approaches for separating inundated class pixels from non-inundated class pixels by utilizing atmospherically corrected S2 data. The experimental results show that alternative approaches are able to achieve high classification accuracy for Camargue and Doñana study areas. Out of the nine alternative approaches tested, three, seven, and four approaches were "Alm Perf" for Camargue, Doñana marshland, and Doñana complete area, respectively. "Alt2 (avg)" and "Alt3 (avg)" provided "Alm Perf" results for both Camargue and Doñana marshland, while "Alt2 (MCET)" and Alt3 (MCET)" provided the most consistent results for all areas, including the Doñana complete area. Thus, "Alt2 (avg)" and "Alt3 (avg)" are suggested for wetlands extensively covered by temporary flooded and emergent vegetation areas, such as the Camargue and the Doñana marshland, while "Alt2 (MCET)" and "Alt3 (MCET)" are expected to give more consistent results for wetlands including a large portion of dry areas, such as the Doñana complete area.

Future steps could consider the exploitation of ancillary information, such as digital elevation models to improve water detection under emergent vegetation, by inferring the water presence based on detected adjacent water covered areas having similar elevation, and land cover information to correct areas erroneously classified as water covered, where water presence is not expected. Furthermore, S2 inundation maps of a site generated via the automatic thresholding alternative achieving top accuracy among other alternatives can be fused with S1 data in order to allow for inundation mapping during extended cloudy periods, based on the example of Reference [12].

Author Contributions: G.A.K. and I.M. conceived and implemented the alternative approaches in this work, which were fine-tuned based on suggestions from G.L. and B.P. G.A.K. performed an accuracy assessment of the generated Sentinel-2 inundation maps. Reference maps for Camargue were provided by G.L. and B.P. Results analysis was carried out by all authors. G.A.K. and I.M. mainly wrote the paper, with significant contribution of G.L. and B.P. to the writing of the discussion and the description of the Camargue study area. All authors reviewed, suggested improvements, and approved the manuscript.

Funding: The European Union's Horizon 2020 research and innovation program under grant agreement No 641762, ECOPOTENTIAL supported and funded this study.

Acknowledgments: We are grateful to Javier Bustamante, Ricardo Díaz-Delgado, and David Aragonés, who are affiliated with the Remote Sensing and GIS Laboratory (LAST-EBD), Estación Biológica de Doñana (CSIC), for kindly providing the reference Landsat inundation maps for Doñana and to Loïc Willm from Tour du Valat Research Institute for providing data and maps from the Camargue.

Conflicts of Interest: The authors declare no conflict of interest.

References

1. World Resources Institute. *Millennium Ecosystem Assessment: Ecosystems and Human Well-Being: Wetlands and Water Synthesis*; World Resources Institute: Washington, DC, USA, 2005.
2. Park, E.; Lee, S.; Peters, D.J. Iowa wetlands outdoor recreation visitors' decision-making process: An extended model of goal-directed behavior. *J. Outdoor Recreat. Tour.* **2017**, *17*, 64–76. [CrossRef]
3. Davidson, N.C. How much wetland has the world lost? Long-term and recent trends in global wetland area. *Mar. Freshw. Res.* **2014**, *65*, 934–941. [CrossRef]
4. Huang, W.; DeVries, B.; Huang, C.; Lang, M.W.; Jones, J.W.; Creed, I.F.; Carroll, M.L. Automated Extraction of Surface Water Extent from Sentinel-1 Data. *Remote Sens.* **2018**, *10*, 797. [CrossRef]
5. Töyrä, J.; Pietroniro, A. Towards operational monitoring of a northern wetland using geomatics-based techniques. *Remote Sens. Environ.* **2005**, *97*, 174–191. [CrossRef]
6. Marti-Cardona, B.; Dolz-Ripolles, J.; Lopez-Martinez, C. Wetland inundation monitoring by the synergistic use of ENVISAT/ASAR imagery and ancilliary spatial data. *Remote Sens. Environ.* **2013**, *139*, 171–184. [CrossRef]

7. Martinis, S.; Plank, S.; Ćwik, K. The use of Sentinel-1 time-series data to improve flood monitoring in arid areas. *Remote Sens.* **2018**, *10*, 583. [CrossRef]
8. Giustarini, L.; Hostache, R.; Matgen, P.; Schumann, G.J.P.; Bates, P.D.; Mason, D.C. A change detection approach to flood mapping in urban areas using TerraSAR-X. *IEEE Trans. Geosci. Remote Sens.* **2012**, *51*, 2417–2430. [CrossRef]
9. Schumann, G.; Di Baldassarre, G.; Bates, P.D. The utility of spaceborne radar to render flood inundation maps based on multialgorithm ensembles. *IEEE Trans. Geosci. Remote Sens.* **2009**, *47*, 2801–2807. [CrossRef]
10. Shen, X.; Wang, D.; Mao, K.; Anagnostou, E.; Hong, Y. Inundation Extent Mapping by Synthetic Aperture Radar: A Review. *Remote Sens.* **2019**, *11*, 879. [CrossRef]
11. Markert, K.N.; Chishtie, F.; Anderson, E.R.; Saah, D.; Griffin, R.E. On the merging of optical and SAR satellite imagery for surface water mapping applications. *Results Phys.* **2018**, *9*, 275–277. [CrossRef]
12. Manakos, I.; Kordelas, G.A.; Marini, K. Fusion of Sentinel-1 data with Sentinel-2 products to overcome non-favourable atmospheric conditions for the delineation of inundation maps. *Eur. J. Remote Sens.* **2019**, 1–14. [CrossRef]
13. Davranche, A.; Poulin, B.; Lefebvre, G. Mapping flooding regimes in Camargue wetlands using seasonal multispectral data. *Remote Sens. Environ.* **2013**, *138*, 165–171. [CrossRef]
14. Donchyts, G.; Schellekens, J.; Winsemius, H.; Eisemann, E.; van de Giesen, N. A 30 m resolution surface water mask including estimation of positional and thematic differences using landsat 8, srtm and openstreetmap: A case study in the Murray-Darling Basin, Australia. *Remote Sens.* **2016**, *8*, 386. [CrossRef]
15. Du, Y.; Zhang, Y.; Ling, F.; Wang, Q.; Li, W.; Li, X. Water bodies' mapping from Sentinel-2 imagery with modified normalized difference water index at 10-m spatial resolution produced by sharpening the SWIR band. *Remote Sens.* **2016**, *8*, 354. [CrossRef]
16. Kyriou, A.; Nikolakopoulos, K. Flood mapping from Sentinel-1 and Landsat-8 data: A case study from river Evros, Greece. In *Proc. SPIE 9644, Earth Resources and Environmental Remote Sensing/GIS Applications VI*; International Society for Optics and Photonics: Toulouse, France, 2015; p. 964405. [CrossRef]
17. Guo, Q.; Pu, R.; Li, J.; Cheng, J. A weighted normalized difference water index for water extraction using Landsat imagery. *Int. J. Remote Sens.* **2017**, *38*, 5430–5445. [CrossRef]
18. Buma, W.; Lee, S.I.; Seo, J. Recent Surface Water Extent of Lake Chad from Multispectral Sensors and GRACE. *Sensors* **2018**, *18*, 2082. [CrossRef]
19. Zhang, F.; Li, J.; Zhang, B.; Shen, Q.; Ye, H.; Wang, S.; Lu, Z. A simple automated dynamic threshold extraction method for the classification of large water bodies from landsat-8 OLI water index images. *Int. J. Remote Sens.* **2018**, *39*, 3429–3451. [CrossRef]
20. Lee, K.S.; Kim, T.H.; Yun, Y.S.; Shin, S.M. Spectral characteristics of shallow turbid water near the shoreline on inter-tidal flat. *Korean J. Remote Sens.* **2001**, *17*, 131–139.
21. Feyisa, G.L.; Meilby, H.; Fensholt, R.; Proud, S.R. Automated Water Extraction Index: A new technique for surface water mapping using Landsat imagery. *Remote Sens. Environ.* **2014**, *140*, 23–35. [CrossRef]
22. Lefebvre, G.; Davranche, A.; Willm, L.; Campagna, J.; Redmond, L.; Merle, C.; Guelmami, A.; Poulin, B. Introducing WIW for detecting the presence of Water In Wetlands with Landsat and Sentinel satellites. *Remote Sens.* **2019**, *11*, 2210. [CrossRef]
23. Kordelas, G.A.; Manakos, I.; Aragonés, D.; Díaz-Delgado, R.; Bustamante, J. Fast and Automatic Data-Driven Thresholding for Inundation Mapping with Sentinel-2 Data. *Remote Sens.* **2018**, *10*, 910. [CrossRef]
24. Díaz-Delgado, R.; Aragonés, D.; Afán, I.; Bustamante, J. Long-term monitoring of the flooding regime and hydroperiod of Doñana marshes with Landsat time series (1974–2014). *Remote Sens.* **2016**, *8*, 775. [CrossRef]
25. Bangira, T.; Alfieri, S.M.; Menenti, M.; van Niekerk, A. Comparing Thresholding with Machine Learning Classifiers for Mapping Complex Water. *Remote Sens.* **2019**, *11*, 1351. [CrossRef]
26. Ban, H.J.; Kwon, Y.J.; Shin, H.; Ryu, H.S.; Hong, S. Flood monitoring using satellite-based RGB composite imagery and refractive index retrieval in visible and near-infrared bands. *Remote Sens.* **2017**, *9*, 313. [CrossRef]
27. Chini, M.; Hostache, R.; Giustarini, L.; Matgen, P. A Hierarchical Split-Based Approach for Parametric Thresholding of SAR Images: Flood Inundation as a Test Case. *IEEE Trans. Geosci. Remote Sens.* **2017**, *55*, 6975–6988. [CrossRef]
28. Otsu, N. A threshold selection method from gray-level histograms. *IEEE Trans. Syst. Man Cybern.* **1979**, *9*, 62–66. [CrossRef]
29. Kittler, J.; Illingworth, J. Minimum error thresholding. *Pattern Recognit.* **1986**, *19*, 41–47. [CrossRef]

30. Ko, B.C.; Kim, H.H.; Nam, J.Y. Classification of potential water bodies using Landsat 8 OLI and a combination of two boosted random forest classifiers. *Sensors* **2015**, *15*, 13763–13777. [CrossRef]
31. DeVries, B.; Huang, C.; Lang, M.; Jones, J.; Huang, W.; Creed, I.; Carroll, M. Automated quantification of surface water inundation in wetlands using optical satellite imagery. *Remote Sens.* **2017**, *9*, 807. [CrossRef]
32. Skakun, S. A neural network approach to flood mapping using satellite imagery. *Comput. Inform.* **2012**, *29*, 1013–1024.
33. Nandi, I.; Srivastava, P.K.; Shah, K. Floodplain mapping through support vector machine and optical/infrared images from landsat 8 OLI/TIRS sensors: Case study from Varanasi. *Water Resour. Manag.* **2017**, *31*, 1157–1171. [CrossRef]
34. Jakovljević, G.; Govedarica, M.; Álvarez-Taboada, F. Waterbody mapping: A comparison of remotely sensed and GIS open data sources. *Int. J. Remote Sens.* **2019**, *40*, 2936–2964. [CrossRef]
35. Miao, Z.; Fu, K.; Sun, H.; Sun, X.; Yan, M. Automatic water-body segmentation from high-resolution satellite images via deep networks. *IEEE Geosci. Remote Sens. Lett.* **2018**, *15*, 602–606. [CrossRef]
36. Isikdogan, F.; Bovik, A.C.; Passalacqua, P. Surface water mapping by deep learning. *IEEE J. Sel. Top. Appl. Earth Obs. Remote Sens.* **2017**, *10*, 4909–4918. [CrossRef]
37. Acharya, T.; Lee, D.; Yang, I.; Lee, J. Identification of water bodies in a Landsat 8 OLI image using a J48 decision tree. *Sensors* **2016**, *16*, 1075. [CrossRef]
38. Yousefi, P.; Jalab, H.A.; Ibrahim, R.W.; Noor, N.F.M.; Ayub, M.N.; Gani, A. Water-body segmentation in satellite imagery applying modified Kernel K-means. *Malays. J. Comput. Sci.* **2018**, *31*, 143–154. [CrossRef]
39. Sivanpillai, R.; Miller, S.N. Improvements in mapping water bodies using ASTER data. *Ecol. Inform.* **2010**, *5*, 73–78. [CrossRef]
40. Jiang, H.; Feng, M.; Zhu, Y.; Lu, N.; Huang, J.; Xiao, T. An automated method for extracting rivers and lakes from Landsat imagery. *Remote Sens.* **2014**, *6*, 5067–5089. [CrossRef]
41. Zhou, Y.; Dong, J.; Xiao, X.; Xiao, T.; Yang, Z.; Zhao, G.; Zou, Z.; Qin, Y. Open Surface Water Mapping Algorithms: A Comparison of Water-Related Spectral Indices and Sensors. *Water* **2017**, *9*, 256. [CrossRef]
42. Haibo, Y.; Zongmin, W.; Hongling, Z.; Yu, G. Water body extraction methods study based on RS and GIS. *Procedia Environ. Sci.* **2011**, *10*, 2619–2624. [CrossRef]
43. Schwatke, C.; Scherer, D.; Dettmering, D. Automated Extraction of Consistent Time-Variable Water Surfaces of Lakes and Reservoirs Based on Landsat and Sentinel-2. *Remote Sens.* **2019**, *11*, 1010. [CrossRef]
44. Chauvelon, P.; Tournoud, M.G.; Sandoz, A. Integrated hydrological modelling of a managed coastal Mediterranean wetland (Rhone delta, France): Initial calibration. *Hydrol. Earth Syst. Sci.* **2003**, *7*, 123–132. [CrossRef]
45. Grillas, P. The Camargue: Rhone River Delta (France). In *The Wetland Book: II: Distribution, Description and Conservation*; Finlayson, C., Milton, G., Prentice, R., Davidson, N., Eds.; Springer: Dordrecht, The Netherlands, 2016.
46. Lefebvre, G.; Germain, C.; Poulin, B. Contribution of rainfall vs. water management to Mediterranean wetland hydrology: Development of an interactive simulation tool to foster adaptation to climate variability. *Environ. Model. Softw.* **2015**, *74*, 39–47. [CrossRef]
47. Lefebvre, G.; Redmond, L.; Germain, C.; Palazzi, E.; Terzago, S.; Willm, L.; Poulin, B. Predicting the vulnerability of seasonally-flooded wetlands to climate change across the Mediterranean Basin. *Sci. Total Environ.* **2019**, *692*, 546–555. [CrossRef] [PubMed]
48. Martí-cardona, B. Spaceborne SAR Imagery for Monitoring the Inundation in the Doñana Wetlands. Ph.D. Dissertation, Polytechnic University of Catalonia, Barcelona, Catalonia, Spain, 2014.
49. Green, A.J.; Bustamante, J.; Janss, G.F.E.; Fernández-Zamudio, R.; Díaz-Paniagua, C. Doñana Wetlands (Spain). In *The Wetland Book: II: Distribution, Description and Conservation*; Finlayson, C., Milton, G., Prentice, R., Davidson, N., Eds.; Springer: Dordrecht, The Netherlands, 2016.
50. Kloskowski, J.; Green, A.J.; Polak, M.; Bustamante, J.; Krogulec, J. Complementary use of natural and artificial wetlands by waterbirds wintering in Doñana, south-west Spain. *Aquat. Conserv. Mar. Freshw. Ecosyst.* **2009**, *19*, 815–826. [CrossRef]
51. Kullback, S. *Information Theory and Statistics*; Dover Publications: New York, NY, USA, 1968.
52. Comaniciu, D.; Meer, P. Mean shift: A robust approach toward feature space analysis. *IEEE Trans. Pattern Anal. Mach. Intell.* **2002**, *24*, 603–619. [CrossRef]
53. Landis, J.R.; Koch, G.G. The measurement of observer agreement for categorical data. *Biometrics* **1977**, *33*, 159–174. [CrossRef]

54. Nakmuenwai, P.; Yamazaki, F.; Liu, W. Automated Extraction of Inundated Areas from Multi-Temporal Dual-Polarization RADARSAT-2 Images of the 2011 Central Thailand Flood. *Remote Sens.* **2017**, *9*, 78. [CrossRef]
55. Martinis, S.; Kuenzer, C.; Wendleder, A.; Huth, J.; Twele, A.; Roth, A.; Dech, S. Comparing four operational SAR-based water and flood detection approaches. *Int. J. Remote Sens.* **2015**, *36*, 3519–3543. [CrossRef]
56. Li, J.; Wang, S. An automatic method for mapping inland surface waterbodies with Radarsat-2 imagery. *Int. J. Remote Sens.* **2015**, *36*, 1367–1384. [CrossRef]
57. Bioresita, F.; Puissant, A.; Stumpf, A.; Malet, J.P. A method for automatic and rapid mapping of water surfaces from sentinel-1 imagery. *Remote Sens.* **2018**, *10*, 217. [CrossRef]
58. Bovolo, F.; Bruzzone, L. A split-based approach to unsupervised change detection in large-size multitemporal images: Application to tsunami-damage assessment. *IEEE Trans. Geosci. Remote Sens.* **2007**, *45*, 1658–1670. [CrossRef]
59. Toral, G.M.; Aragones, D.; Bustamante, J.; Figuerola, J. Using Landsat images to map habitat availability for waterbirds in rice fields. *Ibis* **2011**, *153*, 684–694. [CrossRef]
60. Padró, J.-C.; Pons, X.; Aragonés, D.; Díaz-Delgado, R.; García, D.; Bustamante, J.; Pesquer, L.; Domingo-Marimon, C.; González-Guerrero, Ò.; Cristóbal, J.; et al. Radiometric correction of simultaneously acquired Landsat-7/Landsat-8 and Sentinel-2A imagery using pseudoinvariant areas (PIA): Contributing to the Landsat time series legacy. *Remote Sens.* **2017**, *9*, 1319. [CrossRef]
61. Martins, V.; Barbosa, C.; de Carvalho, L.; Jorge, D.; Lobo, F.; Novo, E. Assessment of atmospheric correction methods for Sentinel-2 MSI images applied to Amazon floodplain lakes. *Remote Sens.* **2017**, *9*, 322. [CrossRef]

 © 2019 by the authors. Licensee MDPI, Basel, Switzerland. This article is an open access article distributed under the terms and conditions of the Creative Commons Attribution (CC BY) license (http://creativecommons.org/licenses/by/4.0/).

Technical Note

The Application of LiDAR Data for the Solar Potential Analysis Based on Urban 3D Model

Iñaki Prieto *, Jose Luis Izkara and Elena Usobiaga

Sustainable Construction Division, 48160 Tecnalia, Spain; joseluis.izkara@tecnalia.com (J.L.I.); elena.usobiaga@tecnalia.com (E.U.)
* Correspondence: inaki.prieto@tecnalia.com; Tel.: +34-607-34-41-03

Received: 1 August 2019; Accepted: 7 October 2019; Published: 10 October 2019

Abstract: Solar maps are becoming a popular resource and are available via the web to help plan investments for the benefits of renewable energy. These maps are especially useful when the results have high accuracy. LiDAR technology currently offers high-resolution data sources that are very suitable for obtaining an urban 3D geometry with high precision. Three-dimensional visualization also offers a more accurate and intuitive perspective of reality than 2D maps. This paper presents a new method for the calculation and visualization of the solar potential of building roofs on an urban 3D model, based on LiDAR data. The paper describes the proposed methodology to (1) calculate the solar potential, (2) generate an urban 3D model, (3) semantize the urban 3D model with different existing and calculated data, and (4) visualize the urban 3D model in a 3D web environment. The urban 3D model is based on the CityGML standard, which offers the ability to consistently combine geometry and semantics and enable the integration of different levels (building and city) in a continuous model. The paper presents the workflow and results of application to the city of Vitoria-Gasteiz in Spain. This paper also shows the potential use of LiDAR data in different domains that can be connected using different technologies and different scales.

Keywords: 3D modelling; LiDAR; CityGML; solar potential

1. Introduction

Currently, half of the world's population lives in urban areas; according to the UN, this number will increase to 60% in two decades. Cities consume a considerable amount of energy, but they can also produce it. Solar energy has the advantage of being able to be generated in the same place it can be consumed due to the possibilities offered by the integration of photovoltaic systems in buildings.

As reflected in Directive 2010/31/EU, 40% of the total energy consumption in the European Union corresponds to buildings. These conditions have caused the EU to promote the development of photovoltaic energy as part of improvement programs for the energy efficiency of buildings. By the end of 2020, at least 25% of new or refurbished buildings will be obliged to comply with the high energy efficiency and bidding requirements for energy consumption, which should be obtained from renewable sources.

Solar energy is the largest and cleanest source of renewable energy. Current technologies enable high performance in the generation of energy from the sun. The potential of solar energy on roofs can be calculated from images, shade estimation, and meteorological data. At the same time, the amount of greenhouse gas emissions avoided in the city with the use of this energy is also estimated [1]. Geographic information systems (GIS) are useful tool for this analysis [2,3].

Work was recently performed based on data obtained with LiDAR technology [4–6]. The accuracy of the results depends on the quality and reliability of the input data. The simplest methods consider the horizontal surface of a roof without taking into account the morphology of a building or the shape of the roof [7]. However, for greater accuracy in the three-dimensional analysis of buildings, other variables, such as the orientation of a roof or even the slope of a roof, must be taken into account without ruling out the effect of shadows [8]. The alternatives presented are based mostly on sophisticated methodologies, commercial tools, and/or complex models.

This paper presents a method for analysing the solar potential of the roofs of urban buildings based on LiDAR data. The method is easily replicable and is based on open data and non-commercial tools that produce high-precision results (1 sqm). In addition, an urban 3D model is generated and semantized with solar potential data for subsequent visualization in a 3D web tool. The integration of information from different domains in a single urban 3D model enables further information retrieval and analysis.

The remaining article is structured as follows: The proposed workflow for solar potential analysis and the urban 3D model creation are explained in Section 2. In Section 3, the workflow is validated in the case study of Vitoria-Gasteiz, Spain. Section 4 contains a discussion of the work, and the main conclusions obtained from the work described in this paper are presented in Section 5.

2. Materials and Methods

To calculate the solar potential and generate the urban 3D model, the following software tools were employed:

- QGIS: GIS open source tool that is used to process layers of geographic information.
- LASTools [9]: powerful LiDAR data-processing tool that is used for data format conversion and the filtering of points that represent the selected urban objects. Figure 1 represents the classification of elements in a LiDAR data file.
- Urban Multi-Scale Environmental Predictor—UMEP Tools [11]: group of environmental services that are implemented as a QGIS plugin. The following services were utilized:

  - UMEP MetPreprocessor: facilitates the adaptation of the EnergyPlus climate file to the meteorological parameters required by the SEBE tool.
  - UMEP Aspect and Height Calculation: calculates the orientations and heights of the facades of buildings from a digital surface model (DSM). Wall aspect is provided in degrees, where a north-facing-wall pixel has a value of zero.

- Solar Energy on Building Envelopes—SEBE Tool [12]: a plugin for QGIS that is used to calculate the pixel-wise potential solar energy using ground and building DSMs.
- CityGML Generation Tool [13]: developed by the authors of this paper to generate a 3D urban model based on the CityGML standard using cadastre, DSM, and digital terrain model (DTM) data [14]. Other tools enable the generation of CityGML models [15].
- The input data used during the process are described as follows:
- Digital Surface Model (DSM): LiDAR file with elevation data of the urban environment, including the elevations of urban elements such as buildings, vegetation or roads.
- Digital Terrain Model (DTM): LiDAR file with elevation data of the ground, on which the urban environment is based (base level of urban elements).
- Weather Data [16]: Detailed climate file of the study area.
- Cadastre Data: GIS file that includes georeferenced dimensions and attributes of land parcels.

Figure 1. LiDAR points classification [10].

Based on these tools and data, Figure 2 shows the process for the analysis of solar generation potential based on an urban 3D model. The workflow is described here.

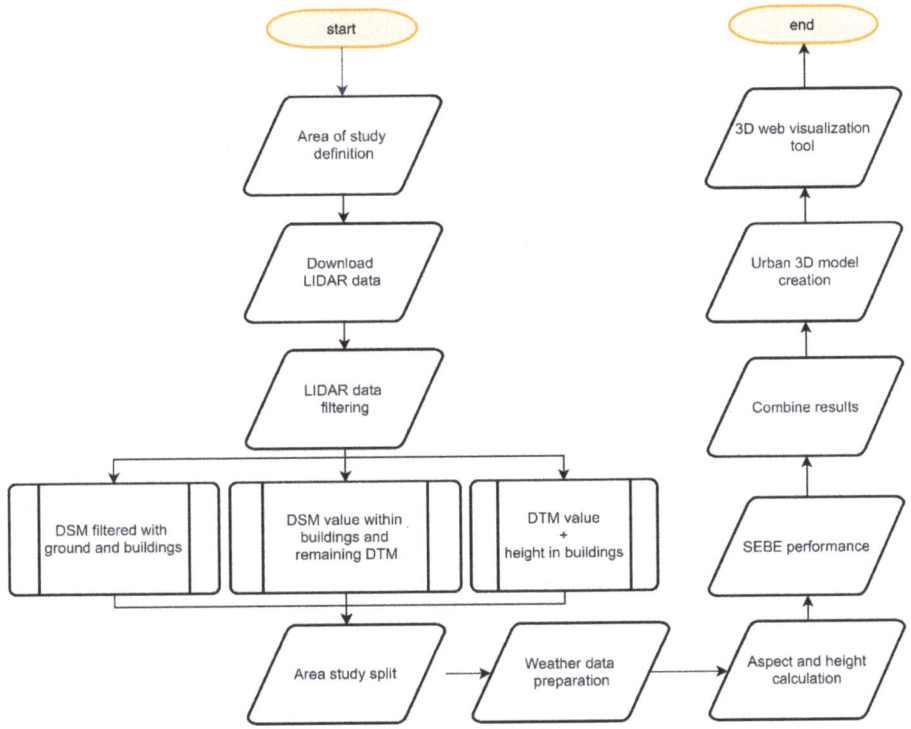

Figure 2. Proposed workflow for solar potential analysis.

The process starts with a definition of the study area (*Area of study definition*). Once the area of study is defined, the required LiDAR files (DSM and/or DTM, depending in the availability) are downloaded (*Download LiDAR data*). The LiDAR files must completely address the area of study. The LiDAR data need to be filtered prior to their usage (*LiDAR data filtering*) using LASTools. As output of this process, a raster file that contains only ground and building points, without vegetation or other objects, is needed. Depending on the availability of the LiDAR data and its quality, three different ways for obtaining this raster file were identified:

- *DSM filtered with ground and buildings*. In the first approach, only DSM LiDAR files are used. These LiDAR files must be in LAS or LAZ format. In this case, the DSM file is filtered by selecting only the points that are classified as 2 (ground) and 6 (building).
- *DSM value within buildings and remaining DTM*. If the DSM classification quality is not as accurate as needed, another approach that uses DTM and DSM data can be used. In this approach, a raster file is created with the DSM values for buildings and DTM values for the surrounding study area.
- *DTM value + height in buildings*. The third approach pertains to cases in which DTM data are available, but DSM data are not available. In addition, we need a building geometry layer (Cadastre GIS file) with a height value as a parameter. In this way, a raster file is created with DTM values (ground) for the study area, with the exception of buildings, for which DTM and building height values are added and a DTM + buildings with flat roof rasters is obtained. Although the quality and precision of this approach are less accurate, a solar potential analysis can be performed in a similar way.

Whichever approach is selected, another step needs to be performed to obtain a complete raster file. Using the QGIS, a geo-process that fills raster regions that lack data values is performed by interpolation from edges. The values for the regions without data are calculated by the surrounding pixel values using inverse distance weighting. Before starting the solar potential analysis, the resulting point cloud for the study area needs to be split into different sections (*Area study split*), which must be rectangular. The creation of a unique raster file to calculate the solar potential is not feasible, as the UMEP tool is not able to perform calculations with such a large raster file. Each section will be independently analyzed using the UMEP tool. The remaining input data required to perform the solar analysis comprises a meteorological file. This file needs to be created in a specific format. The UMEP MetPreprocessor tool enables *Weather data preparation* starting from an EnergyPlus weather file. First, weather data from EnergyPlus [12] are downloaded. From the EnergyPlus weather file, a Comma-Separated Values (CSV) file needs to be created. Second, in the MetPreprocessor tool, a matching between EnergyPlus weather data and UMEP meteorological parameters needs to be defined and performed. Table 1 presents this matching.

Table 1. Matching between EnergyPlus weather file and MetPreprocessor tool.

EnergyPlus Weather Parameter	EnergyPlus Weather Range	UMEP Meteorological Parameter	UMEP Meteorological Range
N1—field Year		Year	
N2—field Month		Month	
N3—field Day		Day	
N4—field Hour		Hour	
N5—field Minute		Minute	
N6—field Dry Bulb Temperature	−70 to 70	Air temperature [°C]	−30 to 55
N8—field Relative Humidity	31,000 to 120,000	Relative humidity	5 to 100
N9—field Atmospheric Station Pressure		Barometric pressure	90 to 107
N13—field Global Horizontal Radiation		Incoming shortwave radiation	0 to 1200
N14—field Direct Normal Radiation		Direct radiation [W m^{-2}]	0 to 1200
N15—field Diffuse Horizontal Radiation		Diffuse radiation [W m^{-2}]	0 to 600
N21—field Wind Speed	0 to 40	Wind speed	0.001 to 60

Last, the UMEP meteorological file, which is subsequently used in the SEBE tool, is obtained. For a detailed calculation of the solar incidence of the building roofs, a prior processing of the raster file of the study area is required to calculate the orientations and heights of the facades of the buildings. This process is performed using the UMEP tool, specifically the *Aspect and height calculation* functionality. This functionality is used to identify the wall pixels and their heights from ground and building digital surface models (DSM) using a filter. The wall aspect can be estimated using a specific linear filter. The wall aspect is given in degrees, where a north-facing-wall pixel has a value of zero. As a result, intermediate files are obtained based on the raster generated for each section of the study area. Intermediate files obtained in this step and the meteorological file that was previously generated are utilized by the SEBE tool to calculate the pixel-wise potential solar energy (*SEBE performance*) using ground and building digital surface models (DSMs). The SEBE calculation needs to be performed for each section of the study area. After the solar potential analysis is performed for all sections, the results are combined in a unique raster layer (*Combine the results*). In addition, the resultant raster layer can be cut with the city geometry to obtain the solar potential for the city limits. The previously calculated solar potential map is bounded to the boundaries of the municipality using the municipality boundary layer. In addition, a radiation threshold was defined for the implementation of solar collection technologies in roofs, in particular 800 kW/m^2 year, and the potential of radiation of the roofs was calculated. As a result of this process, a GIS building layer constructed with solar potential data was obtained. This layer includes the following parameters related to solar potential: (1) useful roof surface (m^2), (2) percentage of useful roof surface (%), (3) total solar radiation (Kwh/year), and (4) solar radiation per sqm (Kwh/m^2·year).

The last step of the process is the generation of a 3D urban model (*Urban 3D model creation*) that incorporates the results obtained from the solar analysis and facilitates the visualization and interpretation of the information contained. The model is based on the CityGML standard defined by the Open Geospatial Consortium (OGC), which combines geometric and semantic information in the same model with different levels of detail. The model generation was performed using the CityGML generation tool. Using DSM and DTM data, the real heights of the buildings are obtained. In this way, 3D buildings can be generated with their real heights and georeferenced, both in position and altitude (on the digital terrain model). As a result, buildings are generated in CityGML LoD2 (refer to Figure 3). The urban 3D model is semantized with the calculated parameters. In this way, all buildings of a city have solar potential analysis values.

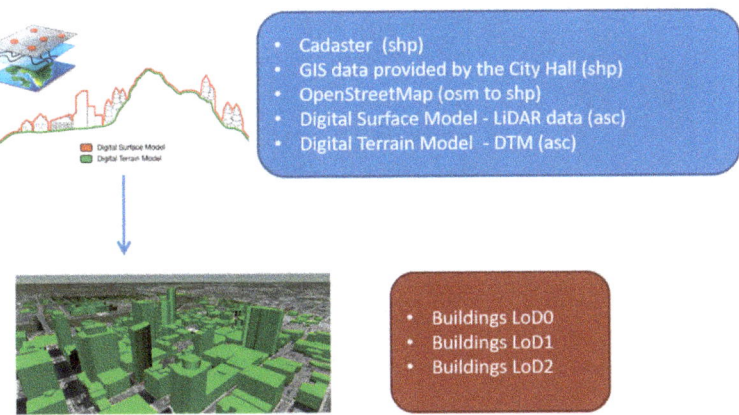

Figure 3. Urban 3D model modelling.

The results are presented in a 3D web tool that enables the visualization of building basic data and solar potential analysis data (*3D web visualization tool*). The information included in the 3D urban model that was previously generated enables the identification of the geographical distribution of the typologies of buildings in the study area. This typological analysis enables the identification of priority areas or districts for solar panel installation, identification of synergies between buildings and adjustment of budget items.

3. Results

In this section, the proposed workflow for the solar potential analysis applied to the Vitoria-Gasteiz case study is presented. The process was performed using the data sources in Table 2.

Table 2. Data sources for solar potential analysis in Vitoria-Gasteiz.

Element	Data Source	Format	Number of Elements
LiDAR	GeoEuskadi [17]	ASC	28 (DSM + DTM)
Buildings	Cadastre [18]	SHP	15.326
Weather	EnergyPlus	EPW	Hourly Data

3.1. Solar Potential Analysis

The selected area of study is the city of Vitoria-Gasteiz in Spain. Sixteen DSM LiDAR files were downloaded for this case study. We performed filtering with ground and buildings points in each file, as the quality of the LiDAR DSM data is sufficient. Three different sections that combine the DSM LiDAR files (as shown in Figure 4) were defined for processing in the UMEP tool.

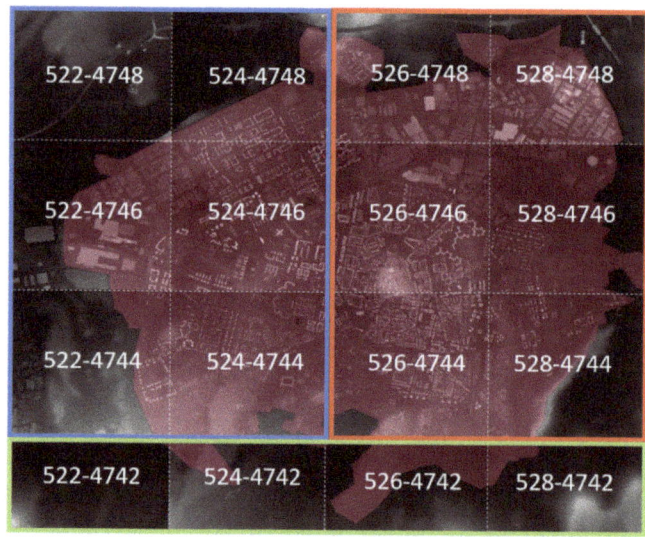

Figure 4. Area study split.

EnergyPlus weather data for Vitoria-Gasteiz was downloaded (Vitoria 080800). These data were processed to obtain UMEP meteorological weather files. The UMEP tool was employed for aspect and height calculations, and the SEBE tool was utilized once for each section, using the same weather file and configuration parameters. The resultant raster layer was combined with the city boundaries to obtain the solar radiation of the study area with a resolution of 1 square metre (refer to Figure 5). The solar radiation map of Vitoria-Gasteiz presents the annual cumulative incident radiation per square meter for roofs in Kwh/m^2·year. The yellow values represent areas of maximum sun exposure, while the blue areas correspond to shadow areas within the city.

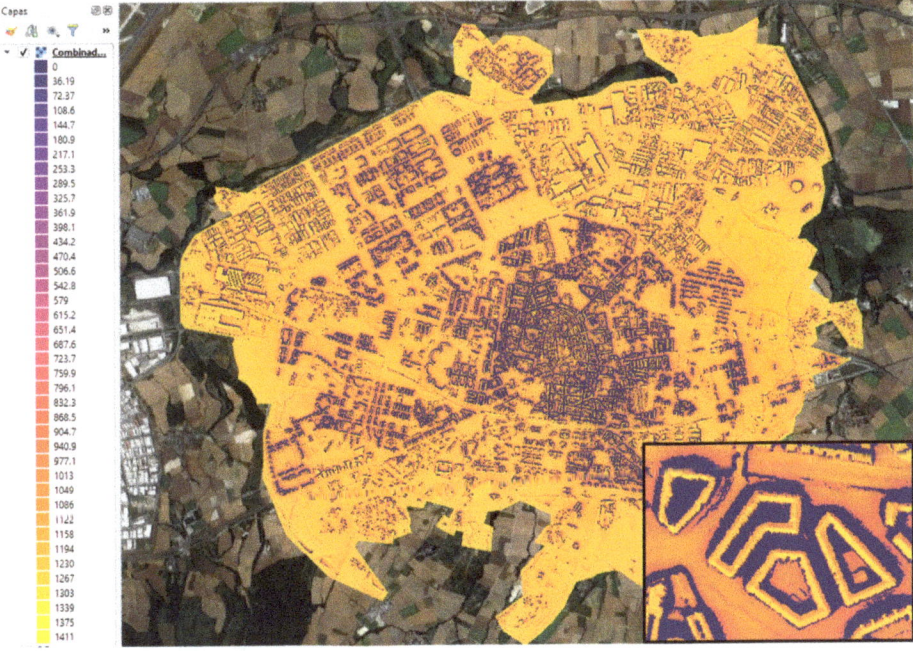

Figure 5. Result of solar potential analysis in Vitoria-Gasteiz (Spain).

3.2. Urban 3D Model Generation

An urban 3D model was generated based on LiDAR (16 DSM and 12 DTM files) and cadastre data presented in Table 2. The urban 3D model was semantized using previously calculated parameters on a building scale (refer to Figure 6). As a result, the following parameters are included in each building in the urban 3D model: (1) gml_id, (2) citygml_measured_height, (3) citygml_measured_height_units, (4) citygml_class, (5) citygml_year_of_construction, (6) citygml_storeys_above_ground, (7) area, (8) rad_total, (9) por_sup_ut, (10) supcub_uti, and (11) rad_m2.

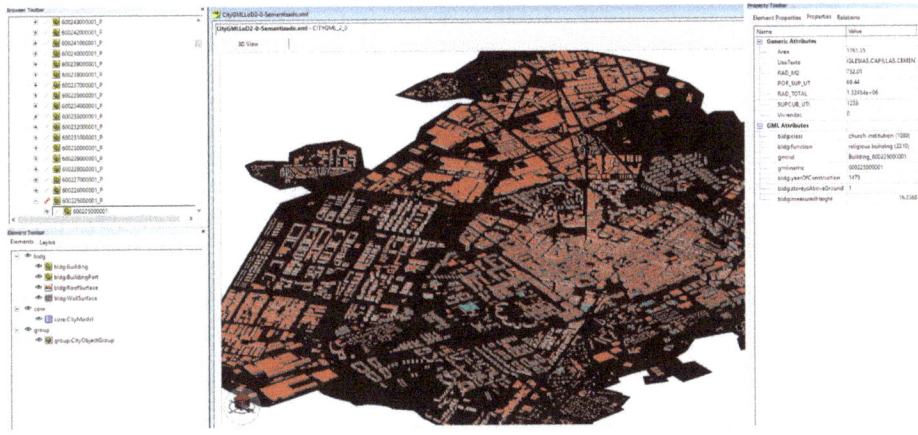

Figure 6. Urban 3D model of Vitoria-Gasteiz (Spain).

3.3. 3D Web Visualization Tool

The 3D web visualization tool integrates a 3D viewer that facilitates the identification and location of buildings in the municipality. For this visualization, the previously generated 3D model is employed. Navigation and interaction are intuitive, as demonstrated in Google Earth, via a 3D map visualization Cesium library. A typological analysis is performed by filters and the combination of several predetermined filters. The visualization of the results is presented through colored maps and statistical data of the results of each type.

The urban 3D model enables a precise and standardized way for the main characteristics of the buildings. The representation of the values of the calculated indicators can be displayed by the 3D viewer for the elements of the model in the study area (refer to Figure 7).

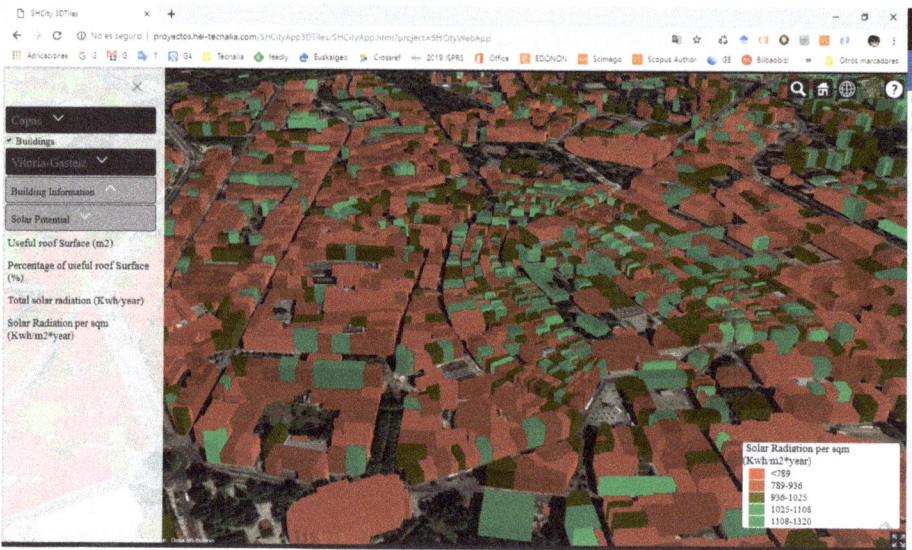

Figure 7. Solar radiation per sqm (Kwh/m^2·year) in Vitoria-Gasteiz (Spain).

4. Discussion

In this section, we discuss the rationale for some of the main decisions made to develop the proposal in this paper.

First, an approach to the solar potential analysis on an urban scale is presented. To calculate the solar potential, we have presented three data input alternatives: DSM with ground buildings; DSM of the building and DTM of the remainder; and DTM + adding height to buildings. The premise is to adapt to different area studies, which usually have different data available. After the analysis of the solar potential in multiple different places, we identified the necessity of systematization in the LiDAR data preparation process to achieve uniformity in the quality and precision.

The results of the study for the city of Vitoria-Gasteiz present values that are similar to the figures offered by the main sources of local and national meteorological data (Basque energy entity—EVE, Spanish National Institute of Meteorology–AEMET). These sources establish the solar radiation incident on the roofs in the city of Vitoria-Gasteiz for a horizontal surface that does not have shadows is 1.390 Kwh/m^2·year, as indicated in the report [19]. This value is very similar to the maximum values obtained using the method proposed in this study.

Second, a 3D city model that is based on the CityGML standard was developed and semantized with all data available on the building level. As a result, a CityGML model is obtained by combining data from different data sources, such as cadastre or solar potential. This model can be subsequently employed as the data layer in different applications, which can involve different agents in the field of municipalities or architects.

As a final advantage of our proposal, previous work (solar potential analysis and the CityGML model) was gathered in a 3D web visualization tool that enables the visualization of the solar potential of each building on the city level.

This research has future limitations that need to be addressed. Adapting the workflow when performing solar potential analysis on large scales (territory). Whether solar potential analysis data can be mapped with existing CityGML ADE, such as CityGML Energy ADE or Solar ADE, should be analyzed.

5. Conclusions

This paper describes the methodology that was followed to perform an analysis of the solar potential-based on LiDAR and the visualization of the results in a 3D web visualization tool. The proposed method is systematic, easily replicable, and based on high-resolution open-data sources and non-commercial software. The results offer high precision and take into account the 3D geometry of buildings, including roof orientation, slope, and the surroundings' orography.

The development of 3D city models that are based on the OGC CityGML standard enables city and building levels to be integrated within a single model that includes both semantic information and geometric information. This model can be used to support multiple applications that different agents, such as urban planners, managers, and citizens, may employ.

The described 3D web visualization tool recognizes the solar potential of each building in the city in a quick, visual, and intuitive way. In addition, the 3D web tool helps to geographically analyze the behaviors of buildings.

The workflow was validated in the city of Vitoria-Gasteiz in Spain. A solar potential analysis was performed, and the urban 3D model was generated and semantized with solar potential data. All gathered data were presented and can be filtered/selected in a 3D web visualization tool.

The results presented in this paper contribute several possibilities for future work. First, the solar potential analysis can be replicated in other municipalities, following the described workflow. Furthermore, the visualization of the results in a 3D web visualization tool eases the interpretation of the data on an urban scale and further information retrieval and analysis.

Author Contributions: I.P. and J.L.I. conceived and implemented the methodology for solar potential calculation. E.U. supported in the data sources compilation and processing. I.P. developed the 3D model and the visualization tool for the case study. I.P and J.L.I. mainly wrote the paper with significant contributions of E.U. in introduction section. All authors reviewed, suggested improvements and approved the manuscript.

Funding: The European Union's Horizon 2020 research and innovation program under grant agreement No 691883, SMARTENCITY supported and funded this study.

Acknowledgments: The work described in this paper was partially funded by SmartEnCity (Towards Smart Zero CO2 Cities across Europe) project, Grant Agreement Number 691883, 2016–2021, as part of the European Union's Horizon 2020 research and innovation program.

Conflicts of Interest: The authors declare no conflict of interest.

References

1. Freitas, S.; Catita, C.; Redweik, P.; Brito, M.C. Modelling solar potential in the urban environment: State-of-the-art review. *Renew. Sustain. Energy Rev.* **2015**, *41*, 915–931. [CrossRef]
2. Gagnon, P.; Margolis, R.; Melius, J.; Phillips, C.; Elmore, R. Estimating rooftop solar technical potential across the US using a combination of GIS-based methods, lidar data, and statistical modeling. *Environ. Res. Lett.* **2018**, *13*, 024027. [CrossRef]
3. Protić, D.D.; Kilibarda, M.S.; Nenković-Riznić, M.D.; Nestorov, I.D. Three-dimensional urban solar potential maps: Case study of the i-scope project. *Therm. Sci.* **2018**, *22*, 663–673. [CrossRef]

4. Martin, A.M.; Dominguez, J.; Amador, J. Applying LIDAR datasets and GIS based model to evaluate solar potential over roofs: A review. *AIMS Energy* **2015**, *3*, 326–343. [CrossRef]
5. Huang, Y.; Chen, Z.; Wu, B.; Chen, L.; Mao, W.; Zhao, F.; Wu, J.; Wu, J.; Yu, B. Estimating roof solar energy potential in the downtown area using a GPU-accelerated solar radiation model and airborne LiDAR data. *Remote Sens.* **2015**, *7*, 17212–17233. [CrossRef]
6. Szabó, S.; Enyedi, P.; Horváth, M.; Kovács, Z.; Burai, P.; Csoknyai, T.; Szabó, G. Automated registration of potential locations for solar energy production with Light Detection and Ranging (LiDAR) and small format photogrammetry. *J. Clean. Prod.* **2016**, *112*, 3820–3829. [CrossRef]
7. Bill, A.; Mohajeri, N.; Scartezzini, J.-L. 3D model for solar energy potential on buildings from urban LiDAR data. In Proceedings of the Eurographics Workshop on Urban Data Modelling and Visualisation, Liège, Belgium, 8 December 2016; pp. 51–56.
8. Kausika, B.B.; Dolla, O.; Folkerts, W.; Siebenga, B.; Hermans, P.; van Sark, W. Bottom-up analysis of the solar photovoltaic potential for a city in the Netherlands: A working model for calculating the potential using high resolution LiDAR data. In Proceedings of the 2015 International Conference on Smart Cities and Green ICT Systems (SMARTGREENS), Lisbon, Portugal, 20–22 May 2015; pp. 1–7.
9. Rapidlasso GmbH LAStools. Available online: https://rapidlasso.com/lastools/ (accessed on 21 August 2019).
10. ArcGIS Lidar Point Classification. Available online: http://desktop.arcgis.com/en/arcmap/10.3/manage-data/las-dataset/lidar-point-classification.htm (accessed on 9 October 2019).
11. Lindberg, F.; Grimmond, C.S.B.; Gabey, A.; Huang, B.; Kent, C.W.; Sun, T.; Theeuwes, N.E.; Järvi, L.; Ward, H.C.; Capel-Timms, I.; et al. Urban Multi-scale Environmental Predictor (UMEP): An integrated tool for city-based climate services. *Environ. Model. Softw.* **2018**, *99*, 70–87. [CrossRef]
12. Lindberg, F.; Jonsson, P.; Honjo, T.; Wästberg, D. Solar energy on building envelopes—3D modelling in a 2D environment. *Sol. Energy* **2015**, *115*, 369–378. [CrossRef]
13. Prieto, I.; Izkara, J.L.; Béjar, R. A continuous deployment-based approach for the collaborative creation, maintenance, testing and deployment of CityGML models. *Int. J. Geogr. Inf. Sci.* **2018**, *32*, 282–301. [CrossRef]
14. Zheng, Y.; Weng, Q.; Zheng, Y. A hybrid approach for three-dimensional building reconstruction in indianapolis from LiDAR data. *Remote Sens.* **2017**, *9*, 310. [CrossRef]
15. Jayaraj, P.; Ramiya, A.M. 3D CityGML building modelling from lidar point cloud data. In Proceedings of the International Archives of the Photogrammetry, Remote Sensing and Spatial Information Sciences—ISPRS Archives, Dehradun, India, 20–23 November 2018.
16. EnergyPlus Weather Data. Available online: https://energyplus.net/weather (accessed on 21 August 2019).
17. GeoEuskadi GeoEuskadi FTP. Available online: ftp://ftp.geo.euskadi.eus/lidar (accessed on 21 August 2019).
18. Tracasa Catastro Alava. Available online: https://catastroalava.tracasa.es/ (accessed on 21 August 2019).
19. Caamaño, E.; Díaz-Palacios, S. Potencial Solar Fotovoltaico de las Cubiertas Edificatorias de la Ciudad de Vitoria-Gasteiz: Caracterización y Análisis. Available online: https://www.vitoria-gasteiz.org/docs/j34/catalogo/01/85/potencialsolar19memoria.pdf (accessed on 21 August 2019).

© 2019 by the authors. Licensee MDPI, Basel, Switzerland. This article is an open access article distributed under the terms and conditions of the Creative Commons Attribution (CC BY) license (http://creativecommons.org/licenses/by/4.0/).

MDPI
St. Alban-Anlage 66
4052 Basel
Switzerland
Tel. +41 61 683 77 34
Fax +41 61 302 89 18
www.mdpi.com

Remote Sensing Editorial Office
E-mail: remotesensing@mdpi.com
www.mdpi.com/journal/remotesensing

www.ingramcontent.com/pod-product-compliance
Lightning Source LLC
LaVergne TN
LVHW070641100526
838202LV00013B/854